14th Anglo-French Physical Acoustics Conference (AFPAC 2015)

Journal of Physics: Conference Series Volume 684

Frejus, France
14 – 16 January 2015

Editors:

Alain Lhemery
Pierre Gelat
Rachel Edwards
Nader Saffari

ISBN: 978-1-5108-2094-4
ISSN: 1742-6588

Printed from e-media with permission by:

Curran Associates, Inc.
57 Morehouse Lane
Red Hook, NY 12571

Some format issues inherent in the e-media version may also appear in this print version.

Printed by Curran Associates, Inc. (2016)

For permission requests, please contact the Institute of Physics
at the address below.

Institute of Physics
Dirac House, Temple Back
Bristol BS1 6BE UK

Phone: 44 1 17 929 7481
Fax: 44 1 17 920 0979

techtracking@iop.org

Additional copies of this publication are available from:

Curran Associates, Inc.
57 Morehouse Lane
Red Hook, NY 12571 USA
Phone: 845-758-0400
Fax: 845-758-2634
Email: curran@proceedings.com
Web: www.proceedings.com

Table of contents

Volume 684

14th Anglo-French Physical Acoustics Conference (AFPAC2015)
14–16 January 2015, Fréjus, France

Accepted papers received: 21 December 2015
Published online: 5 February 2016

Abstract

The Anglo-French Physical Acoustics Conference (AFPAC) had its 14th annual meeting, for the fourth time, at the Villa Clythia, Fréjus, France, from 14th to 16th January 2015. This series of meetings is a collaboration between the Physical Acoustics Group (PAG) of the Institute of Physics and the Groupe d'Acoustique Physique, Sous-marine et UltraSonore (GAPSUS) of the Société Française d'Acoustique.

The Villa manager and his staff worked hard to give attendees the opportunity to spend these three days of intensive work in a very convivial atmosphere.

Four sessions were organized during the three days of the conference, each introduced by an invited talk. As usual (and we could say, as expected), a great variety of subjects were discussed in the 30 presentations that were given, covering various applications and methodologies of physical acoustics, including theoretical and experimental works.

This variety can also be seen when considering the titles of presentations given by our four invited speakers: Dr. Valerie J. Pinfield (Loughborough University), *Ultrasonic propagation in fluid-based dispersions of particles – review, challenges and opportunities*, Dr. Vincent Tournat (Université du Maine) *Wave Processes in Metaporous Materials and Resonant Structures for Improving Sound Absorption*, Prof. Christophe Aristégui (Université de Bordeaux) *Design of soft 3D acoustic metamaterials*, and Prof. I. David Abrahams (Manchester University), *Application of Matched Asymptotic Expansions to Problems in Acoustics*.

The 2014 Bob Chivers prize for the best paper in physical acoustics produced by a PhD student was awarded to Dr. Gim Hwa Chua, for his paper entitled "Use of clicks resembling those of the Atlantic bottlenose dolphin (*Tursiops truncatus*) to improve target discrimination in bubbly water with biased pulse summation sonar IET Radar", published in Sonar and Navigation.

Alain Lhémery (CEA, France), Pierre Gélat (UCL, United Kingdom), Rachel Edwards (University of Warwick, United Kingdom) and Nader Saffari (UCL, United Kingdom).

AFPAC contributes in its own way to the fight against ignorance, mistrust and prejudice.
(press cartoon by Charb, 1967-2015)

Peer review statement

All papers published in this volume of *Journal of Physics: Conference Series* have been peer reviewed through processes administered by the proceedings Editors. Reviews were conducted by expert referees to the professional and scientific standards expected of a proceedings journal published by IOP Publishing.

Crystallographic Orientation Determination of Hexagonal Structure Crystals by Laser Ultrasonic Technique

W Li[1], J Coulson[1], P Marrow[1], R J Smith[1], S J Lainé[2], M Clark[1] and S D Sharples[1]

[1]Applied Optics Group, Electrical Systems and Optics Research Division, Faculty of Engineering, University of Nottingham, University Park, Nottingham, NG7 2RD, UK
[2]Department of Materials Science and Metallurgy, 27 Charles Babbage Road, Cambridge, CB3 0FS, UK

E-mail: wenqi.li@nottingham.ac.uk

Abstract. Spatially resolved acoustic spectroscopy (SRAS) is a laser ultrasonic technique that shows qualitative contrast between grains of different orientation, illustrating the sensitivity of acoustic waves to the material structure. The technique has been improved significantly on determining the full orientation of multigrain cubic metals, by comparing the measured surface acoustic wave (SAW) velocity to a pre-calculated model.

In this paper we demonstrate the ability of this technique to determine the orientation of hexagonal structure crystals, such as magnesium and titanium based alloys. Because of the isotropy of the SAW velocity on the basal plane (0001) of hexagonal crystals, the slowness surface is shown as a circle. As the plane moves from (0001) towards $(11\bar{2}0)$ or towards $(10\bar{1}0)$, the slowness surface gradually turns into an oval. These acoustic properties increase the difficulty in orientation determination.

The orientation results of a grade 1 commercially pure titanium by SRAS is presented, with comparison with electron backscattered diffraction (EBSD) results. Due to the nature of SAWs on hexagonal structure crystals, only the results of Euler angles 1 and 2 are discussed. The error between SRAS and EBSD is also investigated.

1. Introduction

Spatially resolved acoustic spectroscopy (SRAS) is a laser ultrasonic technique for crystallographic orientation determination. It non-destructively measures the surface acoustic wave (SAW) velocity on the material's surface. The microstructure is not only revealed by the velocity contrast, but also the orientation results can be illustrated by the velocities in multiple directions. Although it is complicated to solve the SAW velocity at an arbitrary direction analytically, numerical solutions [1–3], based on the materials elastic constants and the mass density can be determined. We have modified the approach described by Farnell [1] in order to perform the current calculations. The orientation of measured grains can be determined by a brute-force search on this pre-calculated SAWs velocity model as described previously [4].

SRAS has been used for examining cubic crystal structure materials such as nickel-based alloys, aluminium and stainless steel; the SRAS orientation results have been compared to the electron backscattered diffraction (EBSD) technique [5]. The present paper extends previous work to show for the first time how it may be used to obtain quantitative crystallographic orientation for hexagonal structure materials.

In terms of SAWs, the difficulty for hexagonal crystals is that the basal plane (0001) is isotropic; hence, the slowness surface of primary planes $(11\bar{2}0)$ and $(10\bar{1}0)$ are identical. The slowness surface of all planes has both inversion and reflection symmetry. In other words, the hexagonal prism behaves like a cylinder acoustically. Because of this character, on certain planes, where the Euler angle 2 of both planes are the same and both plane normals are in the same plane with axis z as shown in Figure 1, Euler angle 1 can be narrowed down to either 0° or 180° by SRAS. Although only Euler angles 1 and 2 can be determined by SRAS technique, it gives sufficient information for certain applications, for which knowledge of the c-axis direction is of interest.

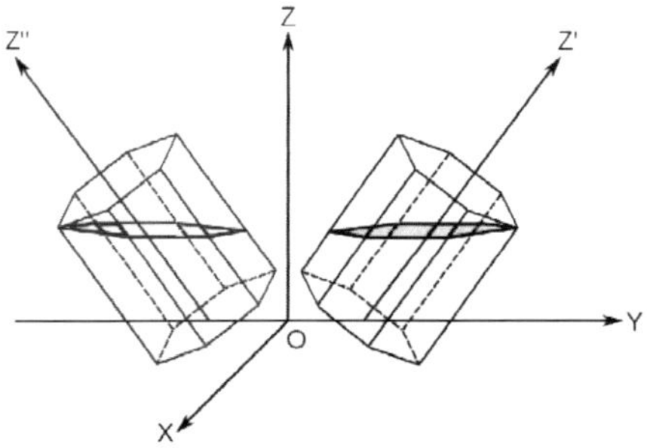

Figure 1: SRAS ambiguity: Euler 1 could be narrowed down to 0° or 180° when axes z, z' and z'' are in the same plane and at the meantime $\angle z''oz \equiv \angle zoz'$.

2. SAW Model of Hexagonal Crystals

The SAW velocity model is calculated based on the material's elastic constants and mass density [4]. An iterative search is performed until the assumed velocity value satisfies the wave equation and the boundary condition of the SAW simultaneously at the known plane and direction. Repeating the calculation on all planes and directions, a numerical model is generated.

Typical planes of titanium α model is shown in Figure 2. The slowness surface is a perfect circle on the basal plane (0001); when the plane gradually moves towards plane $(11\bar{2}0)$ (or plane $(10\bar{1}0)$ as they are acoustically identical), it elongates to an ellipse shape. On some planes, only one wave mode would be detected. Two wave modes could be seen on the planes which are in between the plane (0001) and $(11\bar{2}0)$ (or plane $(10\bar{1}0)$); it is confirmed by the experimental data in later sections.

3. SRAS Instrument

The SRAS technique can be used to obtain SAW velocity information on the local area where the waves are generated. A SAW excitation pattern is generated by projecting a grating pattern of laser light on sample surface. We vary the frequency by using a broadband Nd:YAG laser in combination with a fixed grating [8], allowing the local velocity v can be calculated, through $v = f\lambda$ where f is the frequency and λ is the grating period. By repeating the scan while changing the fringe pattern's direction, a velocity surface spectrum (which can be seen as a 3D slowness surface) of each point on the sample surface is obtained. The current system is capable of scanning at 1200 points/second; the spatial resolution is 100μm, this can be pushed to 80μm for smaller grain size samples. Its operation does not require a vacuum, hence the tested sample size is only limited by the scanning stages' range.

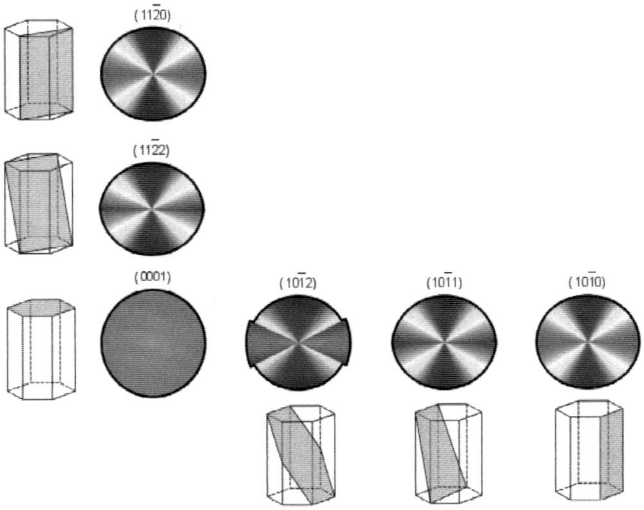

Figure 2: Typical planes in titanium α SAW velocity model. The elastic constants are c11 = 163.60 GPa, c12 = 92.30 GPa, c13 = 67.92 GPa, c33 = 185.20 GPa, c44 = 47.05 GPa [6], and mass density is $\rho = 4.429$ kg/m^3 [7]. The slowness surfaces are shown using both radius and colours, whereby red is a faster velocity and blue is slower. The basal plane (0001) is isotropic in terms of SAW velocity; the slowness surface of plane (11$\bar{2}$0) and (10$\bar{1}$0) are identical acoustically.

4. Orientation Comparison of EBSD and SRAS

A brute-force search, termed the overlap function, was introduced [4] to match the velocity surface spectrum of each point to the pre-calculated SAW model, from which the orientation results are obtained. The inverse pole figures (IPF) on cubic materials were verified by comparison with EBSD technique [5]. This section shows the orientation results of a commercial purity (CP) grade 1 titanium sample.

4.1. Results by the Overlap Function

Figure 3 shown the crystallographic orientation of a CP titanium sample by both EBSD and SRAS. Figure 3a is the IPF measured by EBSD; it is the most common method of presenting materials' texture. A IPF shows the combination of Euler angles 2 and 3, however, due to the limitation of SAWs sensitivity, only Euler angles 1 and 2 can been achieved by SRAS. In order to compare the two techniques visually, a new interpretation for illustrating crystallographic orientation of hexagonal structure materials is introduced, which is as seen in Figure 3b. This method shows the combination of Euler angles 1 and 2, while using three dimensional space of the combination of c-axis pointing direction projected to a colour map. Therefore Figure 3b and Figure 3c are illustrating the same information deduced by EBSD and SRAS respectively. The figure demonstrates the two techniques have a high level of agreement on grains orientation; the few mistakes which occur have been identified for further discussion in the next section.

4.2. Error Analysis

In order to have a better understanding why the overlap function occasionally picked the orientation which is slightly different from EBSD results, experimental data of the grains are shown in Figure 4b and Figure 4d. In addition, the velocity surface spectra are simulated according to the given orientation of EBSD and illustrated in Figure 4a and Figure 4c; they are what SRAS would measure if the orientation is accurate as stated by EBSD results.

Journal of Physics: Conference Series **684** (2016) 012001 doi:10.1088/1742-6596/684/1/012001

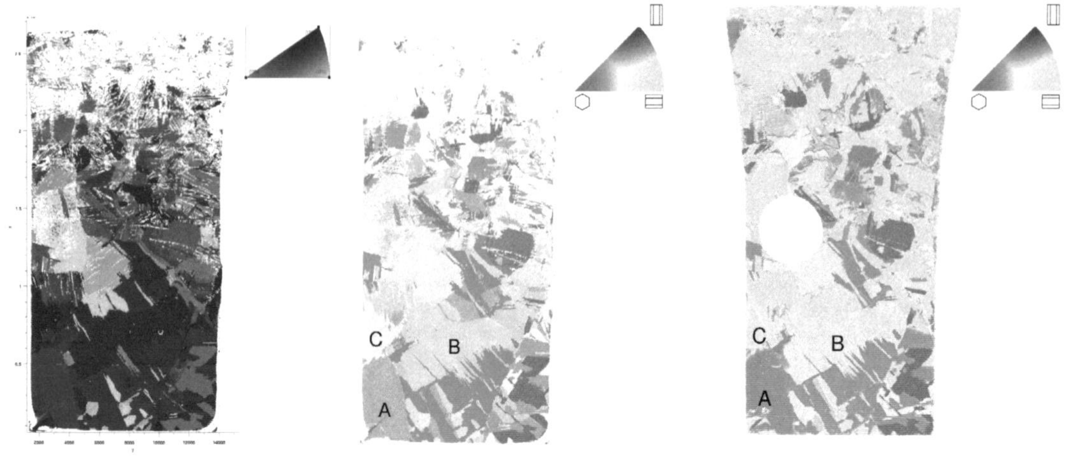

(a) EBSD IPF (Euler angles 2 and 3). Step size is 15 X 15μm.

(b) EBSD c-axis map (Euler angles 1 and 2). Step size is 15 X 15μm.

(c) SRAS c-axis map (Euler angles 1 and 2). Step size is 25 X 25μm.

Figure 3: Titanium orientation results by EBSD (left and middle) and SRAS (right).

Grains	Euler angle 1 (°)		Euler angle 2 (°)	
	EBSD	SRAS	EBSD	SRAS
A	80.19	79.90	49.63	63.96
B	2.71	2.01	76.76	87.91
C	49.01	48.02	68.67	67.98
		50.00		85.99

Table 1: Euler angles 1 and 2 of the three grains measured by EBSD and SRAS.

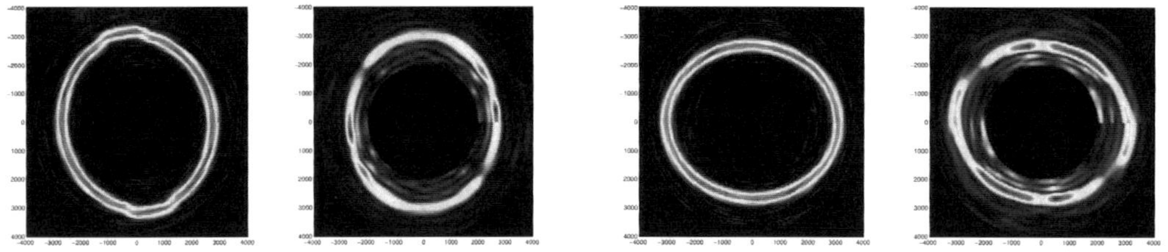

(a) Simulated grain A. (b) Experimental grain A. (c) Simulated grain B. (d) Experimental grain B.

Figure 4: Simulated (from EBSD data) and experimental (measured by SRAS) velocity surface spectra of grains A and B on the CP titanium sample (see Figure 3).

It can be seen in Figure 4, SRAS measured data on some grains are different from what EBSD results show. On grain A, two wave modes should be observed according to EBSD where only one wave mode is measured. On the contrary, two wave modes were detected by SRAS on grain B rather than one, which is indicated by EBSD. Therefore, SRAS and EBSD have minor disagreement on orientation of those grains.

Figure 5 shown the c-axis maps zoom in on grain C by EBSD and SRAS respectively. Figure 5a is the EBSD results which contain Euler angles 1 and 2 shown uniform colour on

Journal of Physics: Conference Series **684** (2016) 012001 doi:10.1088/1742-6596/684/1/012001

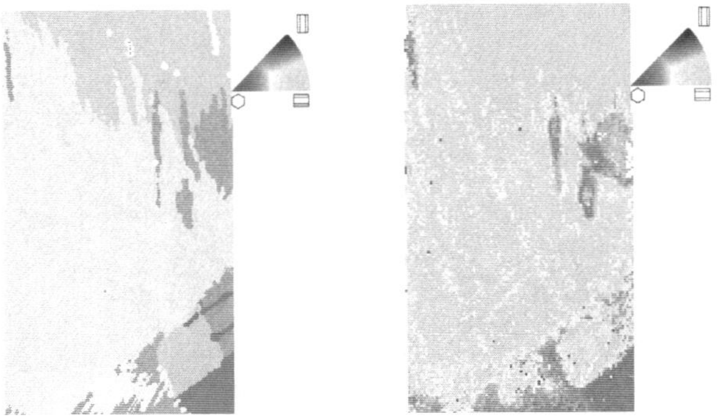

(a) EBSD. (b) SRAS.

Figure 5: The EBSD and SRAS *c*-axis maps zoom in on grain C. The EBSD results pointed out the alpha-phase lath boundaries on Euler angle 3, hence in this figure an uniform colour is shown on this grain while it appeared on the SRAS results.

grain C; the primary alpha laths on this grain can only be seen in Euler angle 3, which should not be observed by SRAS. However, it appeared on the SRAS *c*-axis map in Figure 5b. In Table 1, two sets of data for SRAS as it gives two possible orientations. Euler angle 1 of the two techniques are very close with error less than 1°; one of the SRAS Euler angle 2 is very close to EBSD when the other one is approximately 17° away.

Figure 6a shown the simulated velocity surface spectra from the EBSD data and the measured velocity surface spectra by SRAS (the middle two), of two different areas in yellow and green pixels on SRAS *c*-axis map grain C. The latter are not very easy to distinguish until plotted as slowness surfaces in Figure 6c. The colour of the slowness surfaces match the area on the grain. They are very close in value however one of them has slightly faster velocity in the fastest direction, hence SRAS decides grain C has two different orientations. The disagreement in the results was due to deformation twins being present in the sample, which would mean that there was a second orientation in the grain. The deformation caused different effect on the twinning planes; accordingly, evidence of the third Euler angle is presented. In conclusion, although SRAS has disagreement with EBSD on the orientations of a few grains, it does truly display the elastic properties of the grain's stress/strain conditions.

5. Conclusion

SRAS is a non-destructive evaluation technique to measure the SAW velocity by laser ultrasound. It has proved to be a successful tool for measuring crystallographic orientation quantitatively for cubic structure materials. In this paper CP grade 1 titanium is studied, to show the technique can be extended to the hexagonal structure materials. The calculated SAW velocity model of titanium has shown that its slowness surface is isotropic on the basal plane (0001); when it gradually rotates to plane $(11\bar{2}0)$ or $(10\bar{1}0)$, it elongates into an ellipse shape. The overlap function has been performed to examine the samples' microstructure and orientation of the grains. Although SRAS can only presents two of the Euler angles out of three – physics limits the information that elastic waves can extract of the orientation – the comparison between the *c*-axis maps by EBSD and SRAS confirm it is a robust technique not only for cubic crystals but also hexagonal ones. The mismatches of the two techniques are discussed in the last section. It is believed that the presence of deformation twinning caused the mismatch by SRAS, which indicates this technique is truly representing the elastic properties.

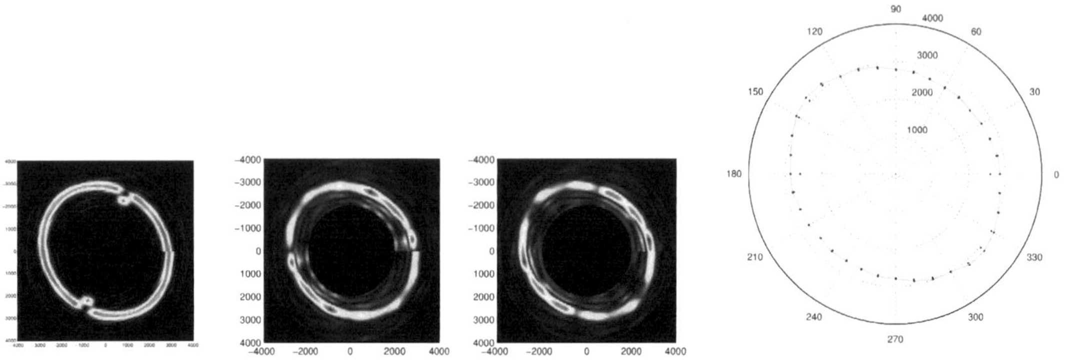

(a) Simulated.

(b) Experimental of the yellow (left) and green (right) pixels.

(c) Experimental slowness surfaces of yellow and green pixels.

Figure 6: The left three figures are the theoretical (simulated from EBSD data) and experimental (measured by SRAS) velocity surface spectra of grains C. The right figure shows the slowness surface of the yellow and green area on grain C by SRAS; the colour of the curves correspond to the pixels' colour.

Acknowledgement
The authors are grateful to Dr Andrew P. Jardine and Stephen M. Walley from the Surfaces, Microstructure and Fracture group at the Cavendish Laboratory, Cambridge for help with the ballistic testing and to Prof D. Rugg from Rolls-Royce plc for the provision of material.

References

[1] G. W. Farnell, *Physical Acoustics - Principles and Methods*, volume VI, chapter 3, 109–166 (Academic Press, New York and London) (1970).

[2] M. Yu. Dvoesherstov, V. I. Cherednik, A. P. Chirimanov 2000 Calculation of Surface Acoustic Waves in Crystals Using the Global Optimization Procedure *Radiophysics and Quantum Electronics* vol 43 issue 9 pp 719-729

[3] Pavla Stoklasov, Petr Sedlk, Hanu Seiner, Michal Landa 2015 Forward and inverse problems for surface acoustic waves in anisotropic media: A RitzRayleigh method based approach *Ultrasonics* vol 56 pp 381-389

[4] Wenqi Li, Steve D Sharples, Richard J Smith, Matt Clark and Michael G Somekh 2012 Determination of crystallographic orientation of large grain metals with surface acoustic waves *Journal of the Acoustical Society of America* vol 132 issue 2 pp 738-745

[5] Wenqi Li, Jethro Coulson, John W Aveson, Richard J Smith, Matt Clark, Michael G Somekh and Steve D Sharples 2013 Orientation Characterisation of Aerospace Materials by Spatially Resolved Acoustic Spectroscopy,`www.ndt.net/article/aero2013/content/papers/31_Li.pdf`

[6] Hirotsugu Ogi, Satoshi Kai, Hassel Ledbetter, Ryuichi Tarumi, Masahiko Hirao, Kazuki Takashima 2004 Titaniums high-temperature elastic constants through the hcpbcc phase transformation *Acta Materialia* vol 52 issue 7 pp 2075-2080

[7] Robert L Mott Machine Elements in Mechanical Design (Fourth Edition), Pearson Education Inc., Upper Saddle River, New Jersey 07458, 2004

[8] Wenqi Li, Steve D Sharples, Matt Clark and Michael G Somekh 2009 Frequency Spectrum Spatially Resolved Acoustic Spectroscopy for Microstructure Imaging *Ultrasonics Symposium (IUS), 2009 IEEE International* pp 791-794

Elastodynamic models for extending GTD to penumbra and finite size flaws

A Kamta Djakou[1], M Darmon[1] and C Potel[2,3]

[1] CEA, LIST, F91191 Gif-sur-Yvette, France
[2] LAUM, UMR CNRS 6613, 72085 Le Mans cedex 9, France
[3] Fédération Acoustique du Nord Ouest (FANO), FR CNRS 3110, France

E-mail: audrey.kamta-djakou@cea.fr

Abstract. The scattering of elastic waves from an obstacle is of great interest in ultrasonic Non Destructive Evaluation (NDE). There exist two main scattering phenomena: specular reflection and diffraction. This paper is especially focused on possible improvements of the Geometrical Theory of Diffraction (GTD), one classical method used for modelling diffraction from scatterer edges. GTD notably presents two important drawbacks: it is theoretically valid for a canonical infinite edge and not for a finite one and presents discontinuities around the direction of specular reflection. In order to address the first drawback, a 3D hybrid method using both GTD and Huygens secondary sources has been developed to deal with finite flaws. ITD (Incremental Theory of Diffraction), a method developed in electromagnetism, has also been developed in elastodynamics to deal with small flaws. Experimental validation of these methods has been performed. As to the second drawback, a GTD uniform correction, the UTD (Uniform Theory of Diffraction) has been developed in the view of designing a generic model able to correctly simulate both specular reflection and diffraction. A comparison has been done between UTD numerical results and UAT (Uniform Asymptotic Theory of Diffraction) which is another uniform solution of GTD.

1. Introduction

The scattering of elastic waves by an obstacle is constituted of phenomena such as the specular reflection and the diffraction. The simplest way to model scattering is the strong approximation of the Geometrical Elastodynamics (GE) which just considers incident and reflected waves. This approximation gives rise to three main regions of space when a plane wave interacts with a crack half-plane: the region illuminated by both incident and reflected waves, the region illuminated only by the incident waves and the shadow region where there is no wave. These regions (see Fig. 1) are separated by shadow boundaries (SB), the incident and reflected shadow boundaries. GE predicts a discontinuity of the elastodynamic field when crossing these shadow boundaries.

This discontinuity of the geometrical field comes from the fact that diffracted waves by the crack edge are not taken into account in GE. There exist a number of methods to model diffraction as the Kirchoff Approximation(KA) [1, 2] and the Geometrical Theory of Diffraction (GTD) [2, 3, 4]. They are all high frequency methods and thus own some limitations. For instance the quantitative prediction by KA of edge diffraction can be erroneous notably for SV waves [5]. Numerical methods as finite elements or finite differences are also used to model the scattering problem. They are time consuming and are generally used for inspection of component or flaws with complex geometry.

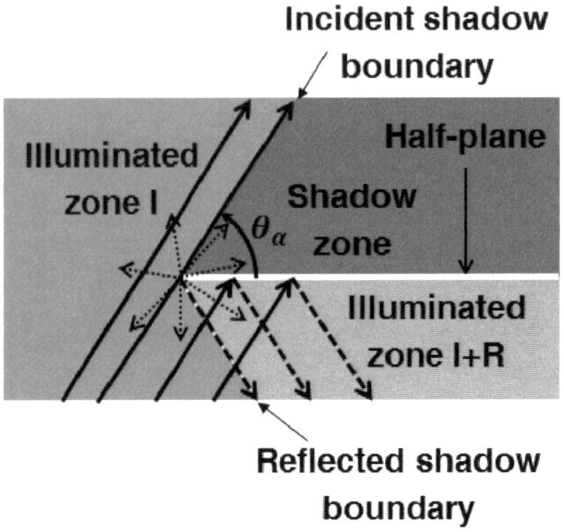

Figure 1: Scattering of a plane wave by a semi-infinite crack embedded in an elastic homogeneous solid. Thick arrows - incident waves, dash arrows - reflected waves, dotted arrows - diffracted waves.

GTD is an extension of GE. Indeed, it adds diffracted rays to incident and reflected rays. These diffracted rays propagate in shadow regions whereby the classical geometrical rays could not (see Fig. 1). The GTD diffracted field is the product of the incident field weighted by a diffraction coefficient and a divergence factor. The phase function of the diffracted field respects the eikonal equation and the divergence factor ensures the energy conservation in a tube of rays. The diffraction coefficient is obtained thanks to canonical configurations as half-planes or wedges. Therefore, GTD does not take into account the finite length of the diffracting edges encountered in NDE. Furthermore, GTD fails in the zones where edge diffracted waves interfere with incident or reflected waves. For this reason, GTD solution is said to be non-uniform.

To overcome these shortcomings of GTD, in a first step, points on the diffracting edge are considered as fictive sources of a field called incremental field. This approach of the problem is called incremental method. Incremental methods have been principally developed in electromagnetism: Incremental Theory of Diffraction (ITD) [6, 7, 8], Incremental Length Diffraction Coefficient (ILDC) [9] and Equivalent Edge Currents (EEC) [10]. They have also been developed in acoustics [11, 12, 13]. In this paper, it is shown that ITD can be extended from electromagnetism to elastodynamics and another incremental model based on Huygens' principle has been developed. In a second step, a uniform extension of GTD, the Uniform Theory of Diffraction (UTD) [14, 15] based on the Pauli-Clemmow process [16], has been developed in elastodynamics in the aim to overcome the second GTD drawback (failure at specular direction). UTD is preferred to UAT (Uniform Asymptotic Theory of diffraction) [17, 18], another uniform extension of GTD, because UAT involves artificial extension of the scattering surface and fictitious reflected rays while UTD does not. UTD is also preferred to the Physical Theory of Diffraction (PTD), a correction of KA in diffraction [19, 20, 21], shown to be equivalent to UAT at the first order [22]. PTD can be time consuming for large scatterers since it is an integral method.

This paper focuses on the development of incremental models and of UTD in elastodynamics. The geometry of the problem is first introduced in section 2. Then, in section 3, the two incremental models, ITD and Huygens-based, are developed in elastodynamics and then experimentally validated. In section 4, UAT result is recalled and UTD is developed in elastodynamics. A numerical comparison between UTD and UAT is performed. Conclusions are provided in section 5.

2. Problem statement

In the following, the symbols α and β are used to denote the wave type, i.e. $\alpha, \beta =$ L, TV or TH (Longitudinal, Transverse Vertical or Transverse Horizontal, respectively). α is used for the incident wave and β for reflected and diffracted waves. Scalar quantities are labelled by taking α and β as subscripts, while α is used as superscripts for vectors related to the incident wave and β as subscripts for vectors related to scattered waves.

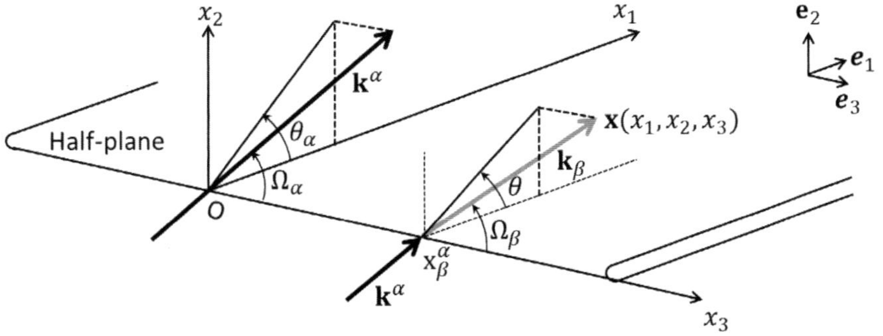

Figure 2: A plane wave of propagation vector $\mathbf{k}^\alpha$ incident on a semi-infinite stress-free crack. Thick black arrow - direction of the incident wave; thick gray arrow - direction of the diffracted wave ($\mathbf{k}_\beta$).

The geometry of the problem is presented in Fig. 2, using the Cartesian system based on an orthonormal basis $\{\mathbf{e_1}, \mathbf{e_2}, \mathbf{e_3}\}$ and the origin O on the crack edge. The crack is embedded in an elastic homogeneous and isotropic space and is the half-plane $\{x_2 = 0, x_1 \geq 0\}$. The edge coincides with the x_3 axis and is irradiated by a plane wave

$$\mathbf{u}^\alpha(\mathbf{x}, \Omega_\alpha, \theta_\alpha) = A\,\mathbf{d}^\alpha e^{i(-\omega t + \mathbf{k}^\alpha \cdot \mathbf{x})}, \tag{1}$$

where A is the wave displacement amplitude; $\mathbf{d}^\alpha$ is its polarization; $\mathbf{k}^\alpha$ its wave vector defined thanks to the angles Ω_α and θ_α (described in Fig. 2), whose magnitude $k_\alpha = \omega / c_\alpha$, with ω - the angular frequency and c_α - the speed of the corresponding mode; t is time; and $\mathbf{x}$ is the position vector, expressed in the Cartesian coordinates as (x_1, x_2, x_3) and in cylindrical coordinates as (r, θ, x_3). The exponential factor $\exp(-i\omega t)$ is implied but omitted everywhere.

The approximate GTD-based total field [4] can be expressed as

$$\mathbf{u}^{\text{tot(GTD)}}(\mathbf{x}, \Omega_\alpha, \theta_\alpha) = \mathbf{u}^{\alpha\text{(GE)}}(\mathbf{x}, \Omega_\alpha, \theta_\alpha) + \sum_\beta \mathbf{u}_\beta^{\alpha\text{(GTD)}}(\mathbf{x}, \Omega_\alpha, \theta_\alpha), \tag{2}$$

where

$$\mathbf{u}^{\alpha\text{(GE)}}(\mathbf{x}, \Omega_\alpha, \theta_\alpha) = H(\eta_\alpha)\,\mathbf{u}^\alpha(\mathbf{x}, \Omega_\alpha, \theta_\alpha) + \sum_\beta H(\eta_\beta)\,\mathbf{u}_\beta^{\alpha\text{(ref)}}(\mathbf{x}, \Omega_\alpha, \theta_\alpha), \tag{3}$$

with H being the Heaviside function and

$$\mathbf{u}_\beta^{\alpha\text{(ref)}}(\mathbf{x}, \Omega_\alpha, \theta_\alpha) = AR_\beta^\alpha\,\mathbf{d}_\beta\,(-q_\beta \cos\theta_\beta)\; e^{i\,k_\beta\,\mathbf{p}_\beta \cdot \mathbf{x}} \tag{4}$$

the reflected field. In equation (4), $\mathbf{p}_\beta = (\sin\Omega_\beta\cos\theta_\beta, -\sin\Omega_\beta\sin\theta_\beta, \cos\Omega_\beta)$ is the unit reflected wave vector and $\mathbf{d}_\beta(-q_\beta\cos\theta_\beta)$ its polarisation vector. The polarisation vector of the reflected field depends on the parameter $-q_\beta\cos\theta_\beta$ where $q_\beta = k_\beta\sin\Omega_\beta$ and θ_β is the reflection angle which is related to the incidence angle θ_α by Snell's law

$$q_\beta\cos\theta_\beta = q_\alpha\cos\theta_\alpha. \tag{5}$$

The diffraction angle Ω_β is related to the incidence angle Ω_α by Snells law in diffraction

$$k_\beta\cos\Omega_\beta = k_\alpha\cos\Omega_\alpha. \tag{6}$$

The arguments $\eta_\alpha = \text{sgn}(\theta - \theta_\alpha)$ and $\eta_\beta = \text{sgn}(\theta - 2\pi + \theta_\beta)$ of the respective Heaviside functions determine whether the observation point is in the illuminated region or shadow of incident and reflected waves (see Fig. 1). The GTD diffracted field [4] is expressed as

$$\mathbf{u}_\beta^{\alpha(\text{GTD})}(\mathbf{x}, \Omega_\alpha, \theta_\alpha) = u^\alpha(x_\beta^\alpha) D_\beta^{\alpha(\text{GTD})}(\theta, \Omega_\alpha, \theta_\alpha)\, \mathbf{d}_\beta(-q_\beta\cos\theta)\, \frac{\mathrm{e}^{ik_\beta S_\beta}}{\sqrt{k_\beta L_\beta}}, \tag{7}$$

where $u^\alpha(x_\beta^\alpha) = \mathbf{u}^\alpha(\mathbf{x}_\beta^\alpha)\cdot\mathbf{d}^\alpha$, $x_\beta^\alpha(0, 0, x_3 - S_\beta\cos\Omega_\beta)$ is the diffraction point on the scattering edge (see Fig. 2); $D_\beta^{\alpha(\text{GTD})}$ is the so-called diffraction coefficient; S_β is the distance between the diffraction point $\mathbf{x}_\beta^\alpha$ and the observation point $\mathbf{x}$; $L_\beta = S_\beta\sin^2\Omega_\beta$ is a distance parameter and $\mathbf{d}_\beta(-q_\beta\cos\theta)$ is the diffracted polarisation vector. In (7), $k_\beta L_\beta$ is the far field parameter. GTD diffraction coefficients contain poles $\theta = \theta_\beta$ and $\theta = 2\pi - \theta_\beta$. These poles correspond respectively to the directions of incident and reflected shadow boundaries (see Fig. 1). GTD solution thus diverges at shadow boundaries and is obtained for canonical configurations such as the half-plane in this case. It does not take into account the finite length of the obstacle.

3. Incremental methods

At the diffraction point x_β^α, any crack of contour $\mathcal{L}$ is locally approximated by a canonical form, a half-plane in this case (see Fig. 3). This half-plane is tangent to the crack at the diffraction point. The diffraction point x_β^α is the origin of the orthonormal basis vector $\{\mathbf{e_1}, \mathbf{e_2}, \mathbf{e_3}\}$ associated to the half-plane. In this localised basis, (S, ϕ, θ) are the spherical coordinates of the observation vector $\mathbf{x}$.

Incremental methods assume that points of the diffracting edge are virtual sources of a field defined as the incremental field $\mathbf{F}_\beta(x_\beta^\alpha, \mathbf{x})$. Thus, the field diffracted by the contour $\mathcal{L}$ at an observation point is the integral of the incremental field on the contour $\mathcal{L}$:

$$\mathbf{u}_\beta^\alpha(\mathbf{x}, \Omega_\alpha, \theta_\alpha) = \int_{\mathcal{L}} \mathbf{F}_\beta(x_\beta^\alpha, \mathbf{x})\,\mathrm{d}l, \tag{8}$$

with $\mathrm{d}l$ being the edge increment. We have developed two different methods in elastodynamics to determine this incremental field: one based on the GTD locality principle (ITD) and one based on the Huygens principle.

3.1. ITD in elastodynamics

ITD is based on the locality principle. It has been developed in electromagnetism [6] and has been extended in elastodynamics in [23]. The ITD incremental field is

$$\mathbf{F}_\beta(x_\beta^\alpha, \mathbf{x}) = \frac{u^\alpha(x_\beta^\alpha)}{\sqrt{2i\pi}} \sin\phi\, D_\beta^{\alpha(\text{GTD})}(\theta, \Omega_\alpha(\phi), \theta_\alpha) \frac{e^{ik_\beta S}}{S}\, \mathbf{d}_\beta(-q_\beta\cos\theta), \tag{9}$$

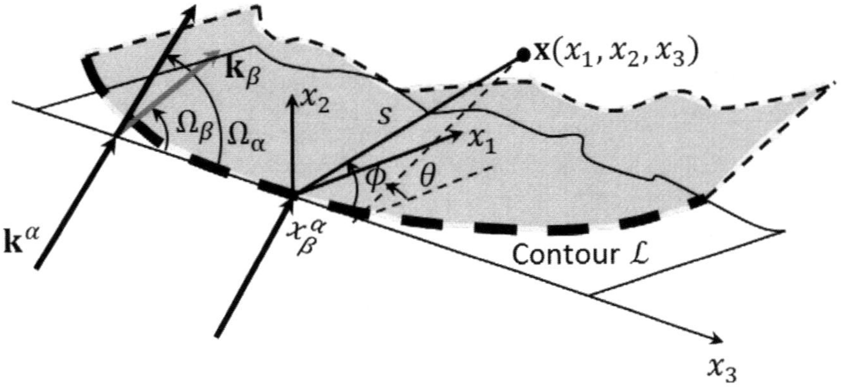

Figure 3: A plane wave of propagation vector $\mathbf{k}^\alpha$ incident on a stress-free obstacle (in gray) of contour $\mathcal{L}$. Thick black arrow - direction of the incident wave; thick gray arrow - direction of the wave $\mathbf{k}_\beta$ scattered by the half-plane tangent to the crack edge at the flash point x^α_β.

where

$$\Omega_\alpha(\phi) = \arccos\left(\frac{k_\beta}{k_\alpha}\cos\phi\right). \tag{10}$$

This incremental field, valid in the far field zone $k_\beta S \gg 1$, is a spherical wave weighted by a coefficient. Therefore, the contour points act as fictive source of spherical waves. It has been checked that ITD gives back GTD solution in the case of the plane wave scattering from a half-plane i.e. for an infinite straight edge. Another incremental model based on the Huygens principle is developed in next section.

3.2. Huygens method

The second method is based on the Huygens principle. It also supposes that points on the diffracting edge are fictive sources of spherical waves which interfere between each other and give rise to the displacement field at an observation point. An Ansatz is proposed so that each spherical wave is weighted by a diffraction coefficient proportional to GTD diffraction coefficient $D^{\alpha(\mathrm{GTD})}_\beta$. The diffracted field can therefore be written as

$$\mathbf{u}^\alpha_\beta(\mathbf{x}, \Omega_\alpha, \theta_\alpha) = \int_{\mathcal{L}} \mathrm{C}\, u^\alpha(x^\alpha_\beta) D^{\alpha(\mathrm{GTD})}_\beta(\theta, \Omega_\alpha, \theta_\alpha)\frac{e^{ik_\beta S}}{S}\,\mathbf{d}_\beta\left(-q_\beta\cos\theta\right) dl \tag{11}$$

where $\mathcal{L}$ is the crack contour; C is an unknown parameter to determine and dl is the edge increment. To find the value of the parameter C, the stationary phase method is applied to (11) in the case of a straight infinite edge and equalized to (7) because the contribution of the stationary point corresponds to the edge diffracted field. Finally,

$$\mathrm{C} = \frac{\sin\Omega_\beta}{\sqrt{2i\pi}}. \tag{12}$$

The incremental fields ITD and Huygens-based are respectively

$$\mathbf{F}_\beta(x^\alpha_\beta, \mathbf{x})\big|_{\mathrm{ITD}} = u^\alpha(x^\alpha_\beta)K(\phi(l))\,\frac{e^{ik_\beta S}}{S}\,\mathbf{d}_\beta\left(-q_\beta\cos\theta\right), \tag{13}$$

$$\mathbf{F}_\beta(x^\alpha_\beta, \mathbf{x})\big|_{\mathrm{Huygens}} = u^\alpha(x^\alpha_\beta)K(\Omega_\beta)\,\frac{e^{ik_\beta S}}{S}\,\mathbf{d}_\beta\left(-q_\beta\cos\theta\right), \tag{14}$$

with the coefficient

$$K(\zeta) = \frac{\sin\zeta}{\sqrt{2i\pi}} D_{\beta}^{\alpha(\mathrm{GTD})}(\theta, \Omega_{\alpha}(\zeta), \theta_{\alpha}). \tag{15}$$

Huygens method (14) differs from ITD (13) in the case of a straight edge only by the argument ζ of the coefficient $K(\zeta)$. For Huygens, this argument ζ is the diffraction angle Ω_β linked to the incidence angle Ω_α by Snell's law of diffraction (6) whereas for ITD it is the angle ϕ between the observation point and the discretization point (see Fig. 3). These incremental models can be applied to GTD and also to uniform GTD corrections. They have been validated experimentally in the following.

3.3. Experimental validation

Diffracted echoes generated by the top tip of a 40 mm large and 10 mm high backwall breaking planar notch simulated by incremental methods are compared to experimental results for various flaw orientations with respect to the probes one.

Diffraction echoes have been measured in the TOFD (Time Of Flight Diffraction) inspection of a ferritic steel component (see Fig. 4) using two 6.35 mm diameter mono-element probes emitting P45 waves at 2.25 MHz with a 60 mm PCS (Probe Centre Spacing). This ferritic steel component is defined in the Cartesian coordinates (X, Y, Z) shown in Fig. 4. The defect is initially perpendicular to X−axis so that it is inspected in a 2D configuration. Then, the skew angle (angle between the flaw top edge and the Y-axis) is increased from 0 ° (2D configuration) to 70 ° (see Fig. 4) in order to be in a 3D configuration by rotating the specimen around the Z-axis. The echo generated by a 2 mm diameter and 40 mm length side-drilled hole (see Fig. 4) is employed for calibration.

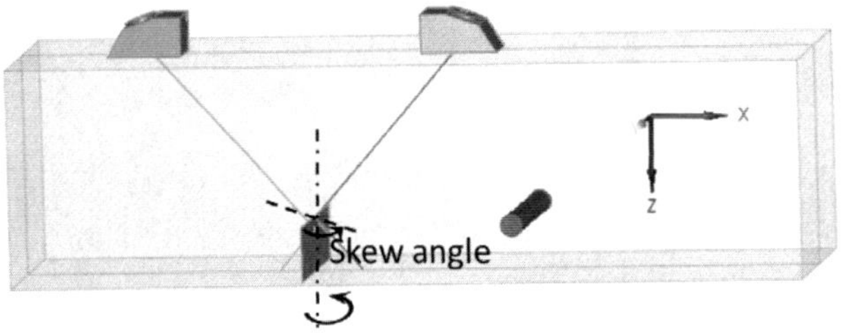

Figure 4: TOFD simulation configuration.

The measured and simulated results are presented in Table 1. ITD and Huygens lead to the same results. The errors between ITD/Huygens simulations and experimental results are at most or around 1dB and are less than the measurements incertitudes (around 2dB).

4. Uniform corrections of GTD in elastodynamics

The classical edge-diffracted GTD ray field is not valid near and at shadow boundaries (directions of specular reflection and direct transmission). Indeed GTD evaluates asymptotically the exact solution of the scattering from a half-plane which is a Sommerfeld integral [24, Ch. 3] and just takes into account the contribution of the integral stationary phase point. This contribution corresponds to the diffracted field whereas the integrand's poles contribution gives rise to the geometrical field. To handle the coalescence of stationary phase points and integrand's poles, interference of diffracted waves with incident and reflected waves of GE, other methods of approximation are used such as the Van Der Waerden's one which gives rise to UAT and the Pauli-Clemmow's one which leads rise to UTD. UAT has already been derived in elastodynamics by [4, 18].

Table 1: Amplitude of diffracted waves in dB.

Skew angle ($^\circ$)	experiments	ITD	Huygens
0	-13.1	-12.9	-12.9
10	-12.6	-13.0	-13.0
20	-14.2	-13.1	-13.1
30	-13.3	-13.3	-13.3
40	-13.7	-13.7	-13.7
50	-13.6	-14.3	-14.3
60	-14.0	-15.2	-15.2
70	-14.4	-16.3	-16.2

4.1. UAT

The approximate UAT-based total field [4, 17, 18] can be expressed as

$$\begin{aligned} \mathbf{u}^{\mathrm{tot(UAT)}}(\mathbf{x}) &= A\left[\overline{F}(\xi_\alpha) - \widehat{\overline{F}}(\xi_\alpha)\right] e^{i\mathbf{k}^\alpha\cdot\mathbf{x}}\,\mathbf{d}^\alpha \\ &\quad + \sum_\beta AR^\alpha_\beta \left[\overline{F}(\xi_\beta) - \widehat{\overline{F}}(\xi_\beta)\right] e^{ik_\beta\,\mathbf{p}_\beta\cdot\mathbf{x}}\,\mathbf{d}_\beta\,(q_\beta\cos\theta_\beta) \qquad (16) \\ &\quad + \sum_\beta u^\alpha(\mathbf{x}^\alpha_\beta) D^{\alpha(GTD)}_\beta \frac{e^{ik_\beta S_\beta}}{\sqrt{k_\beta L_\beta}}\,\mathbf{d}_\beta\,(q_\beta\cos\theta)\,, \qquad (17) \end{aligned}$$

where

$$\widehat{\overline{F}}(X) = e^{i\frac{\pi}{4}}\frac{e^{iX^2}}{2X\sqrt{\pi}}, \qquad (18)$$

$$\overline{F}(X) = \frac{1}{\sqrt{i\pi}}\int_X^{+\infty} e^{it^2}\,dt \qquad (19)$$

is the Fresnel function and the parameters

$$\xi_\alpha = -\mathrm{sgn}\,(\theta-\theta_\alpha)\sqrt{2k_\alpha L_\alpha\sin^2\left(\frac{\theta_\alpha-\theta}{2}\right)} \qquad (20)$$

$$\text{and}\quad \xi_\beta = -\mathrm{sgn}\,(\theta+\theta_\beta-2\pi)\sqrt{2k_\beta L_\beta\sin^2\left(\frac{\theta_\beta+\theta}{2}\right)} \qquad (21)$$

are the detour parameters [17]. The divergence of GTD near shadow boundaries is removed by the function $\widehat{\overline{F}}$. The asymptotic development of the Fresnel function for large arguments [25] is

$$\overline{F}(X) \simeq H(-X) + \widehat{\overline{F}}(X). \qquad (22)$$

Then, far away of the shadow boundaries, UAT total field (16) gives back GTD total field (2). In this UAT solution, the incident and reflected field are no more defined only in illuminated regions. They are defined on the whole space. Indeed, UAT needs to extend reflected rays in the shadow region by introducing fictitious rays [26].

4.2. UTD
The approximate UTD-based total field [27] can be expressed as

$$\mathbf{u}^{\mathrm{tot(UTD)}}(\mathbf{x},\Omega_\alpha,\theta_\alpha) = \mathbf{u}_\beta^{\alpha(\mathrm{GE})}(\mathbf{x},\Omega_\alpha,\theta_\alpha) + \sum_\beta F(k_\beta L_\beta\, a)\,\mathbf{u}_\beta^{\alpha(\mathrm{GTD})}(\mathbf{x},\Omega_\alpha,\theta_\alpha), \qquad (23)$$

where F is a transition function and the parameter a describes the proximity of the observation point to a shadow boundary. When the observation point is far away from the shadow boundaries, the transition function tends to 1 and then, UTD is equal to GTD. When the observation point is close to the shadow boundaries, the transition function tends to 0 and removes the singularity of GTD diffraction coefficient; it also introduces a discontinuity in the diffracted field at the shadow boundaries which is cancelled by the GE one so that the total UTD field is continuous at shadow boundaries contrary to GTD.

The UTD just modified the amplitude of the diffracted rays. It does not require fictitious rays as UAT. It is then more simple to implement. UTD asymptotics do not include all terms of the order $(k_\beta L_\beta)^{-1/2}$ as UAT [26, 28]. A comparison between UAT and UTD is performed in the next section.

4.3. Numerical results
The numerical results are presented in the $(\mathbf{e}_1, \mathbf{e}_2)$ plane, which is perpendicular to the edge crack, since the problem is invariant along the x_3 direction (see Fig. 2). The observation vector $\mathbf{x}$ is specified using the cylindrical coordinates (r, θ) associated with the $(\mathbf{e}_1, \mathbf{e}_2)$ plane. The solid material used for simulations is ferritic steel with Poisson's ratio $\nu = 0.29$, longitudinal velocity $c_L = 5900$ m.s^{-1} and transversal velocity $c_T = 3230$ m.s^{-1}.

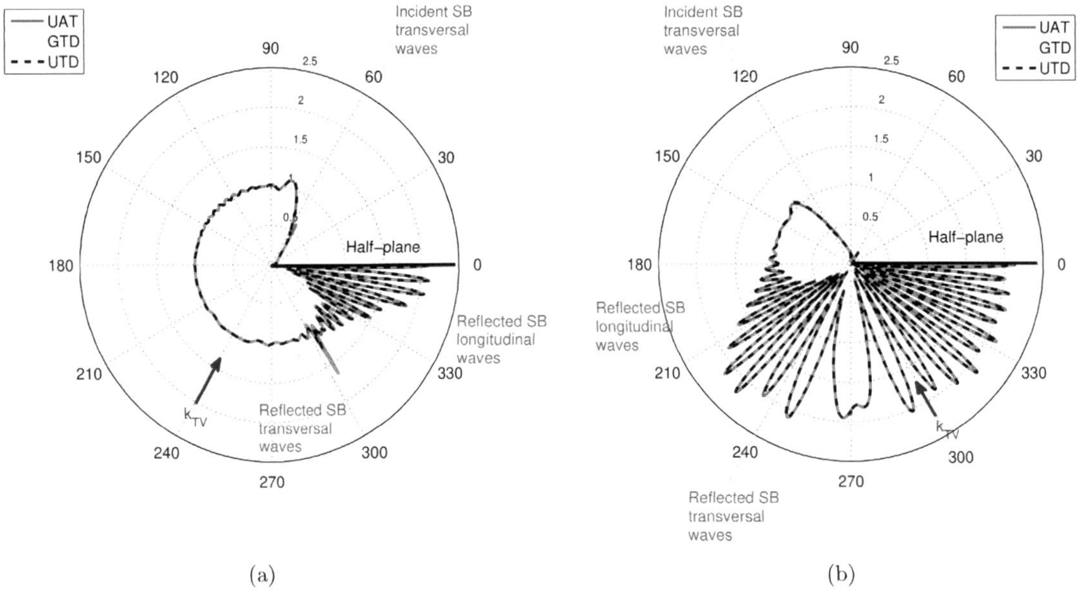

Figure 5: Directivity pattern of the total field predicted by different models (GTD, UAT and UTD) at $r = 10\lambda\,(k_\beta L_\beta = 20\pi,\ \beta = TV)$ for $\Omega_{TV} = 90^\circ$ - (a) $\theta_{TV} = 60^\circ$ - (b) $\theta_{TV} = 120^\circ$. Each circle represents amplitude of the total field normalized by the incident amplitude.

In Fig. 5, all approximate total fields are calculated using all scattered modes, with GTD-based non uniform asymptotics given by (2), UAT - by (16), and UTD - by (23). The used configuration is 2D: the incidence is normal to the edge crack ($\Omega_\alpha = 90^\circ$). The incident wave is transverse vertical. There are three shadow boundaries in Fig. 5.a, the incident TV shadow boundary $\theta = 60^\circ$, reflected TV shadow boundary $\theta = 300^\circ$ and reflected L shadow boundary $\theta \approx 336^\circ$. There also have three shadow boundaries in Fig. 5.b, the incident TV shadow boundary $\theta = 120^\circ$, reflected TV shadow boundary $\theta = 240^\circ$ and reflected L shadow boundary $\theta \approx 204^\circ$. The peaks near critical angles $\theta_c \approx 56.8^\circ$ and $2\pi - \theta_c \approx 303,2^\circ$ are due to the interference of the diffracted T wave with the corresponding head waves generated by diffraction at the edge (see Fig. 6).

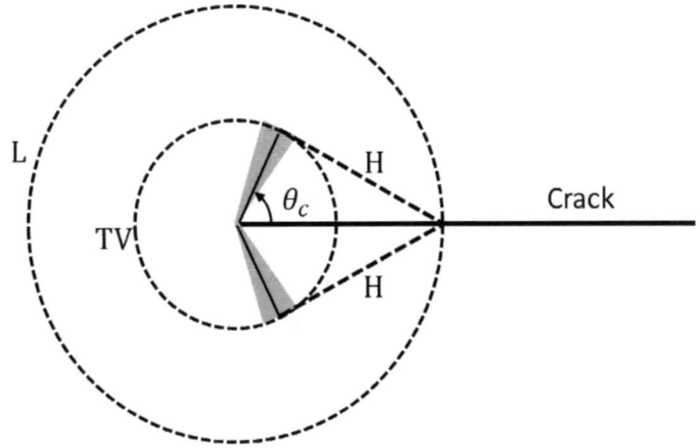

Figure 6: Waves diffracted from a crack edge: L stands for longitudinal wave, TV for transversal and H for head wave; θ_c is the longitudinal critical angle.

As expected, both UTD and UAT are continuous at shadow boundaries and practically coincide far away from the shadow boundaries. There is a slight difference between them near the shadow boundaries. In Fig 5.a, there is a difference of peak amplitude at the critical angles between UAT and UTD. That is because, such critical angles are located near the incident and reflected shadow boundaries. In this observation region the diffracted rays (contribution of stationary phase point) interfere with incident or reflected rays (poles) and head waves (branch point). UAT reproduces the peaks of GTD field since it just modifies the geometrical field whereas in UTD, the transition function tends to 0 near shadow boundaries and then reduces the head wave peak amplitude. Nevertheless, both GTD and its uniform corrections are classically not considered to be valid for observation near a critical angle [29]. A modified version of UTD is proposed in [27] so that its prediction is the same as GTD or UAT at critical angle θ_c for a longitudinal incident wave.

In Fig 5.b, the poles are far away from the critical angles. In this case, UTD and UAT give approximately the same result far and near shadow boundaries. The maximum error in dB between UTD and UAT is 2dB.

5. Conclusion

This paper focuses on recent advances in modelling the scattering of elastic waves from an obstacle. The GTD ray method, classically used to model edge diffraction, is valid for an infinite edge and not for a finite size scatterer. Moreover, GTD diverges at observation directions of specular reflection and direct transmission (called shadow boundaries). To overcome these limitations of GTD, two incremental models and a GTD uniform correction have been derived in elastodynamics. The two developed ITD and Huygens' models permit to take into account the finite size of the scatterer edges. These two models give back GTD solution in the case of a straight infinite edge. They have also been both successfully validated experimentally. UTD, a uniform correction of GTD, permits to have a continuous total field. It modifies the amplitude of the diffracted rays near the shadow boundaries in order to remove the GTD divergences and to cancel the discontinuity of the geometrical field. UTD is simpler to implement than UAT which required to build fictitious rays. It gives results close to UAT ones. Incremental methods can be coupled with UTD to build a generic model both spatially uniform and able to deal with finite size flaws.

References

[1] R. K. Chapman, "Ultrasonic scattering from smooth flat cracks: An elastodynamic Kirchhoff diffraction theory," CEGB Report, North Western Region NDT Applications Centre, NWR/SSD/82/0059/R, 1982.
[2] M. Darmon, N. Leymarie, S. Chatillon and S. Mahaut, "Modelling of scattering of ultrasounds by flaws for NDT." Ultrasonic wave propagation in non homogeneous media. Springer Berlin Heidelberg, 61–71 (2009).
[3] J. B. Keller, "Geometrical theory of diffraction," *J. Opt. Soc. Am.* **52**, 116–130 (1962).
[4] J. D. Achenbach, A.K. Gautesen and H. McMaken, *Rays methods for waves in elastic solids* (Pitman, New York, 1982).
[5] M. Darmon, V. Dorval, A. Kamta Djakou, L. Fradkin and S. Chatillon, "A system model for ultrasonic NDT based on the Physical Theory of Diffraction (PTD)," *Ultrasonics*, **64**, 115-127 (2016).
[6] R. Tiberio and S. Maci, "An incremental theory of diffraction: scalar formulation," *IEEE Trans. Ant. Prop.*, **42**, 600–612 (1994).
[7] R. Tiberio, S. Maci and A. Toccafondi, "An incremental theory of diffraction: Electromagnetic formulation," *IEEE Trans. Ant. Prop.*, **43**, 87–96 (1995).
[8] R. Tiberio, A. Toccafondi, A. Polemi and S. Maci, "Incremental theory of diffraction: a new-improved formulation," *IEEE Trans. Ant. Prop.*, **52**, 2234–2243 (2004).
[9] A. D. Yaghjian, "Incremental length diffraction coefficients for arbitrary cylindrical scatterers," *IEEE Trans. Ant. Prop.*, **49**, 1025–1032 (2001).
[10] Y. Z. Umul, "Rigorous expressions for the equivalent edge currents," *Progress in Electromagnetism Research B*, **15**, 77–94 (2009).
[11] U. P. Svensson, R. I. Fred and J. Vanderkooy, "An analytic secondary source model of edge diffraction impulse responses," *J. Acoust. Soc. Am.*, **106**, 2331–2344 (1999).
[12] U. P. Svensson, P. T. Calamia and S. Nakamishi, "Frequency-domain edge diffraction for finite and infinite edges," *Acta. Acust. Acust.*, **95**, 568–572 (2009).
[13] T. K. Stanton, Dezhang Chu and G. V. Norton, "Acoustic diffraction by deformed edges of finite length: theory and experiment," *J. Acoust. Soc. Am.*, **122**, 3167–3176 (2007).
[14] P. H. Pathak and R. G. Kouyoumjian, "The dyadic diffraction coefficient for a perfectly-conducting wedge," DTIC Document, Tech. Rep. (1970).
[15] R. G. Kouyoumjian and P. H. Pathak, "A uniform geometrical theory of diffraction for an edge in a perfectly conducting surface," *Proc. of the IEEE* **62**, 1448–1461 (1974).
[16] P. C. Clemmow, "Some extension to the method of integration by steepest descent," *Quart. Journ. Mech. and Appl. Math.* **III**, 241–256 (1950).
[17] S. W. Lee and G. A. Deschamps, "A uniform asymptotic theory of electromagnetic diffraction by a curved wedge," *IEEE Trans. Ant. Prop.* **24**, 25–34 (1976).
[18] H. McMaken, "A uniform theory of diffraction for elastic solids," *J. Acoust. Soc. Am.* **75**, 1352–1359 (1984).
[19] P. Ya Ufimtsev, *Fundamentals of the physical theory of diffraction* (John Wiley & Sons, New Jersey, 2007).
[20] V. A. Borovikov and B. Y. Kinber, *Geometrical theory of diffraction* (The institution of Electrical Engineers, 1994).
[21] B. Lü, M. Darmon, C. Potel, L. Fradkin and V. Zernov, "Models Comparison for the scattering of an acoustic wave on immersed targets," *J. Phys. Conf. Series* **353**, doi: 10.1088/1742-6596/353/1/012009 (2012).
[22] V. Zernov, L. Fradkin and M. Darmon, "A refinement of the Kirchhoff approximation to the scattered elastic fields". Ultrasonics, **52**, 830–835 (2012).
[23] A. Kamta Djakou, M. Darmon and C. Potel. "Elastodynamic models for extending GTD to finite length defects". Submitted (2015).
[24] V. M. Babich, M. A. Lyalinov and V. E. Grikurov *Diffraction theory, the Sommerfeld-Malyuzhinets technique* (Alpha Sience, 2008), Ch. 3.
[25] V. A. Borovikov, *Uniform stationary phase method* (The institution of Electrical Engineers, 1994).
[26] D. Bouche and F. Molinet, *Méthodes asymptotiques en électromagnétisme*, (Springer-Verlag, 1994).

[27] A. Kamta Djakou, M. Darmon, L. Fradkin and C. Potel, "The Uniform geometrical Theory of Diffraction for elastodynamics: plane wave scattering from a half-plane". To be published in JASA (2015).
[28] F. Molinet, *Acoustic high-frequency diffraction theory,* (Momentum press, New York, 2011).
[29] L. Fradkin, M. Darmon, S. Chatillon and P. Calmon, "A semi-numerical model for the near-critical angle scattering." Submitted (2015).

Thin-film transducers for the detection and imaging of Brillouin oscillations in transmission on cultured cells

F Pérez-Cota[1], R J Smith[1], E Moradi[2], K Webb[2] and M Clark[1]

[1]Electrical Systems and Optics Research Division, University of Nottingham, University Park, Nottingham, UK

[2]Institute of Biophysics, Imaging and Optical Science (IBIOS), University of Nottingham, University Park, Nottingham, UK

E-mail: eexfap@nottingham.ac.uk

Abstract. Mechanical imaging and characterisation of biological cells has been a subject of interest for the last twenty years. Ultrasonic imaging based on the scanning acoustic microscope (SAM) and mechanical probing have been extensively reported. Large acoustic attenuation at high frequencies and the use of conventional piezo-electric transducers limit the operational frequency of a SAM. This limitation results in lower resolution compared to an optical microscope. Direct mechanical probing in the form of applied stress by contacting probes causes stress to cells and exhibits poor depth resolution. More recently, laser ultrasound has been reported to detect ultrasound in the GHz range via Brillouin oscillations on biological cells. This technique offers a promising new high resolution acoustic cell imaging technique. In this work, we propose, design and apply a thin-film based opto-acoustic transducer for the detection in transmission of Brillouin oscillations on cells. The transducer is used to generate acoustic waves, protect the cells from laser radiation and enhance signal-to-noise ratio (SNR). Experimental traces are presented in water films as well as images of the Brillouin frequency of phantom and fixed 3T3 fibroblast cells.

1. Introduction

Mechanical characterisation of biological cells is of great interest, as their properties (like stiffness or elasticity) are mostly unknown and can be the basis for mechanical modelling. Moreover, it provides an alternative mechanism for contrast which could provide new insights into cell biology. This kind of characterisation can be performed on a cell by applying stress and measure the deformations directly, with for example, an atomic force microscope (AFM) tip [1], by optical trapping [3] or micropipette aspiration [4]. Those approaches have the advantage that measurements are obtained directly. However, these are invasive techniques which can easily stress or damage a cell. Alternatively, ultrasound has been used to perform such measurements. Using a scanning acoustic microscope (SAM), mechanical images of living cells have been reported [5] as well as maps of the speed of sound [6]. Ultrasound is less invasive and features can be resolved beyond the surface. The SAM output depends on two or more parameters meaning that the quantitative measurements are not obtained directly and its resolution is typically lower than that of optical systems.

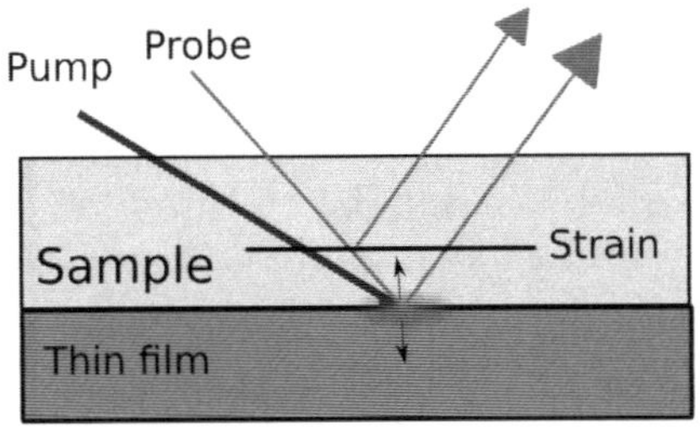

Figure 1: Schematic of Brillouin scattering detection. The light scattered at the acoustic wave interferes with the reference beam reflected from the metal film producing Brillouin oscillations.

Ultrasound is a promising way of high frequency non-destructive imaging. The speed of sound being significantly slower than the speed of light makes the acoustic wavelength comparable to the optical wavelength when ultrasound is in the GHz range. However, conventional methods to generate ultrasound are limited by the high attenuation of sound at GHz frequencies. Also, electronics and piezoelectric transducers become increasingly difficult to make in this regime. Frequencies used in a SAM to image biological cells are of the order of 1GHz and having typical wavelength of approximately $1.5\mu m$ which is lower than that used on an optical microscope.

Laser ultrasound offers a way to generate and detect high frequency ultrasound [7]. It is a non-contact, wireless, couplant-free technique capable of producing acoustic waves of very high frequency (up to THz [8]). Also, the pump-probe technique [9], it provides a mean to detect very high frequencies without the use of special detectors or electronics. Recently, picosecond laser ultrasound measurements have been performed on single biological cells [10]. The detection method presented in [10] is based on Brillouin scattering which provides a direct measure of the speed of sound and the refractive index on translucent materials. In this method, the beam scattered by a GHz acoustic wave is combined with a reference which causes an intensity modulation called Brillouin oscillations (see figure 1). The frequency of these oscillations is know as the Brillouin frequency and, for normal incidence, it is given by the relation:

$$f_B = \frac{2n\nu}{\lambda} \tag{1}$$

where ν is the speed of sound, n the refractive index and λ the optical wavelength. If the refractive index is known, the speed of sound, which is related to the mechanical properties, can be easily calculated. Moreover, the interrogated acoustic wavelength is half of the optical one used to probe the ultrasound. This technique has being demonstrated on vegetal [10], bone [11] or HeLa cells[12]. A shift in the Brillouin frequency from cytoplasm to nucleus was found indicating good contrast within a cell. The Brillouin frequency has also been used to measure the speed of sound and elasticity of biological cells[11]. However, this technique has not been applied for imaging which could be a step towards high-resolution acoustic microscopy. The main limitations to perform such imaging are given by the slow acquisition speed of the pump-probe technique, direct exposure of the cell to laser beams and poor signal to noise ratio (SNR). Low SNR is due to several factors; the limitation of used power due to the sensitivity of a cell to laser light, high losses in the detection path of a pump-probe system and the distortion of laser spots caused by light beams travelling through a heterogeneous scattering object(the cell itself).

We present a three-layer thin-film transducer designed for the detection of Brillouin scattering in transmission. The transducer serves as a shield to reduce laser light exposure while enhancing SNR. Moreover, the acquisition speed is increased by the use of a pump-probe ASOPS system [13]. These allowed to perform imaging of the Brillouin frequency of phantom and fixed cells.

2. Design of thin-film transducers for the detection of Brillouin scattering in transmission

Thin film transducers are used in this work for the generation of acoustic waves. The transducers consist of three thin layers of gold, indium tin oxide (ITO) and gold (see figure 4). The transducers were originally designed to improve the sensitivity to compression waves in picosecond laser ultrasound experiments [2] and have been re-engineered for the detection of Brillouin scattering in transmission. The wavelengths used in our pump-probe system, for generation (λ_{pump}=390nm) and detection (λ_{probe}=780nm) of the acoustic waves, govern the choice of materials; the gold films strongly absorb at λ_{pump} while reflecting/transmitting at λ_{probe} allowing flexibility of design.

Detection in transmission is the main feature of the transducers presented here since it protects the cells from laser light while increasing SNR to obtain images. This is done by exposing the lasers to the transducer through a glass coverslip and detecting the probe after it is transmitted through the sample. This increases detection levels and reduces loss of spot quality. Moreover, it gives the opportunity to increase laser power without compromising the integrity of the cells.

For that to be possible, it is necessary to optically model the transducers in order to optimise transmission (for detection) and absorption(for generation). Additionally, when the transducer is exposed to a short light pulse, it resonates not only optically but acoustically as well. The frequency of this mechanical resonance, which depends on the characteristics of the layers, has implications on the design. In this section, optical and mechanical models will be presented for the design and characterisation of the transducers.

2.1. Optical characterisation

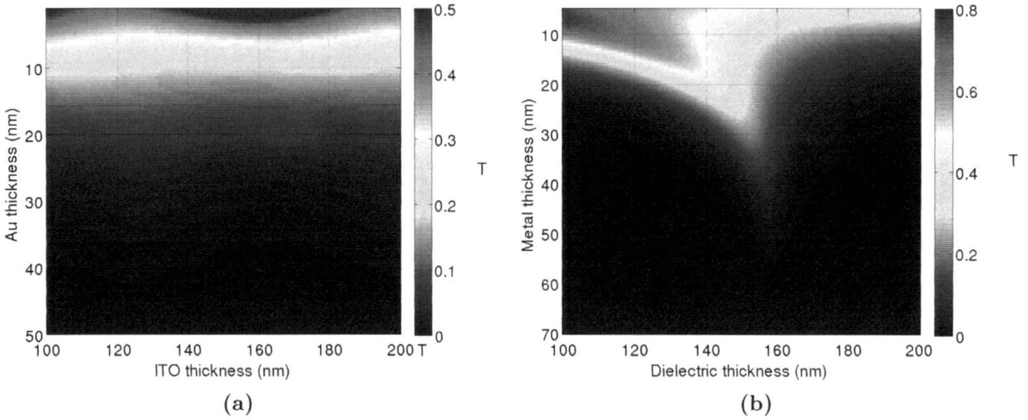

Figure 2: Simulation of the optical transmittance of the transducers for gold layers from 5-70nm and ITO layers from 100-200nm. (a) The transmittance for λ_{pump}. (b) The transmittance for λ_{probe}.

The transducers consist of an optical cavity with partially reflecting layers similar to a low order Fabry-Perot interferometer. When the transducer is exposed to light, part of the beam enters the cavity and it is resonantly transmitted and reflected. The transducer has a sensitivity to incoming waves that depend on the wavelength and dimensions of the layers [2]. Based on Fresnel coefficients, a model was created with the assumption of infinite width of the plate which is a fair approximation given the width is greater than the optical wavelength. The model

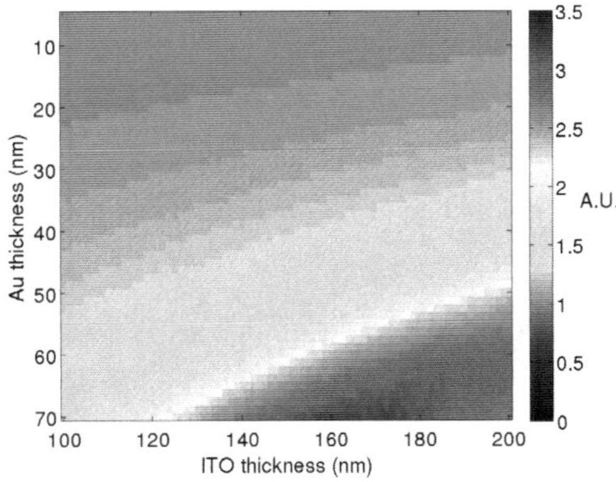

Figure 3: Simulation of the amplitude generated by the transducer at ~5.5GHz (approximately the Brillouin frequency for a cell at λ_{probe}). The amplitude shows little change in the zone transducers are expected to be built (~20nm of gold and ~140nm of ITO). The increase in amplitude observed for the thicker layers is due to the resonance of the transducer shifting down to 5.5GHz.

calculates transmission at λ_{probe} and λ_{pump} assuming identical gold layer thicknesses. Figure 2 shows the result of the model. From figure 2a it is possible to see that the transmittance at λ_{pump} shows little change against ITO which means all the absorption occurs within the gold layers. Figure 2b shows that transmittance of λ_{probe} experiences a resonant peak as the ITO layer changes allowing high transmittance for relatively thick gold layers.

2.2. Mechanical Characterisation

Using finite element modelling (FEM), the transducers were modelled using a coupled optical-thermal-mechanical one dimensional model to predict the displacement resultant from optical pulse excitation. The output of the model, which consists of the displacement of the layers, is used to calculate the main resonance frequency as well as the generated amplitude at the expected Brillouin frequency from a cell (~5.5GHz). The natural resonant frequency of the device, which is expected to be detected, changes from 6-18GHz over the range presented in figure 3 and ~9-11GHz for the dimensions the transducers are expected to be built(~20nm of gold and ~140nm of ITO) [2]. Figure 3 shows a simulation of the generated amplitude at 5.5GHz for the same layer thicknesses presented in figure 2. A small variation in the generated amplitude is observed for transducers with gold layers below 50nm. For devices with gold layers above 50nm, the amplitude increases significantly because the resonant frequency of the transducer shifts down closer to 5.5GHz.

2.3. Transmission detection and design criteria

The transducers are intended to be used to protect cells against laser radiation while enhancing signal-to-noise-ratio. Figure 4 shows the proposed detection method. Lasers are incident on the transducer through the glass substrate keeping laser spots from distorting. Ultrasound is optically generated while the cell remains protected from pump radiation (beam not shown for clarity) which allow extended exposure for imaging. This transmission configuration also simplifies the detection path since only a few elements are required to capture the probing light. This simplification reduces the detection path losses in comparison with a conventional reflection mode path where polarising optics are used to separate beams travelling in opposite directions.

The design of the transducers is a compromise between optical and acoustical performance. From figure 2a, we can see that above 20nm thickness of gold, the pump transmittance at λ_{pump} is ~0.1 which ensures a reduction of the amount of light the cell is exposed to while absorption is high for acoustical generation (not shown). Thicker gold layers would absorb even more pump light, however transmittance of λ_{probe} would be compromised. Figure 2b shows that the transmittance at λ_{probe} can be up to 0.3 for gold layers of approximately 20nm if the thicknesses

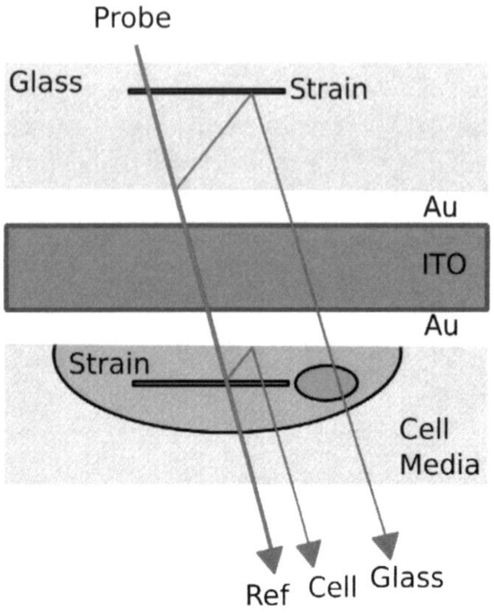

Figure 4: Schematic of the detection of Brillouin scattering in transmission based on thin-film transducers. This approach helps reduce the exposure of the cell to laser light while increasing SNR. Due to high transmittance of the film, two Brillouin signals are expected one scattered from the sample(ref+cell) and another one from the substrate(ref+glass).

of ITO is selected round the peak in transmission(~140nm).

Mechanically, thicker gold layers can increase the amplitude of the generated wave at 5.5GHz (see figure 3). This condition could increase the amplitude of the detected signals in our experiments. Unfortunately, is not possible to take advantage of this effect at our current probing wavelength since the the thicknesses required to increase the generated amplitude (gold layers greater than 50nm) are not suitable for high optical transmission (see figure 2). Moreover. to match the generated acoustic frequency with the aimed Brillouin frequency (~5.5GHz) brings a practical problem as becomes difficult to separate those signals from the acquired time traces.

The sound waves generated on the transducer travel through the sample and glass substrate and since the transmittance is high, meaning a second Brillouin signal from the glass is observed. As the glass substrate is considerably stiffer than most cells, the detected Brillouin frequency from the glass substrate can be easily separated from the rest of the signal by digital filtering.

Taking into consideration the models shown in this section, a transducer with 20 +/-5nm and 145+/-15nm of gold and ITO layer thicknesses respectively, is suitable for detection of Brillouin scattering in transmission. The device will have a resonating frequency of approximately 9-11Ghz [2] which is easy to separate, in the frequency domain, from the Brillouin signal. A 0.25-0.4 transmittance for λ_{probe} allows a good detection level with a reduced input power at the sample (that depends on experimental set-up). For λ_{pump} an approximate absorption of 0.4 means good light to sound conversion and a 0.1 transmittance means reduced exposure to the pump beam that biological cells are most sensitive to.

Reduction of the intensity that the cells are exposed to at λ_{pump} allows a raise in pump power, a reduction of the number of averages and an increment of the number of measurements. That leads to greater SNR, faster acquisition speed and the acquisition of an image. At the same time, the transducer transmits a portion of the probe beam to allows grater detection levels compared to a conventional reflection mode set-up. The configuration proposed here is particularly useful in the case of the sample being a biological cell, where the delicate subjects are easily affected by laser light.

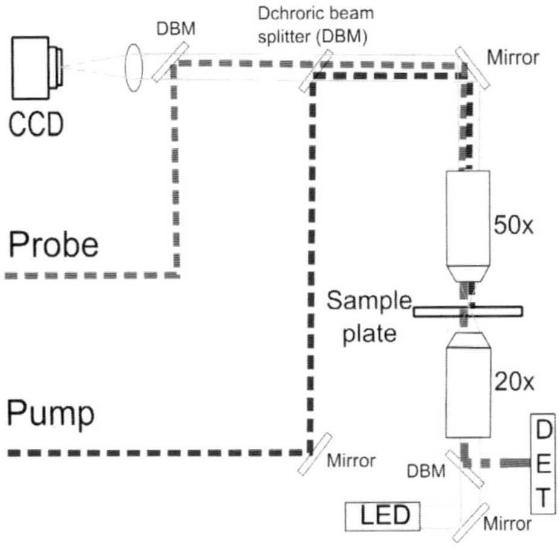

Figure 5: Schematic of the Experimental set-up. Two pulsed lasers paths are combined by dichroic beam splitters and focused to the sample with a 50x long working distance objective. The probe beam is captured by a 20x objective. Optical imaging of the sample is done in parallel to the acoustic imaging.

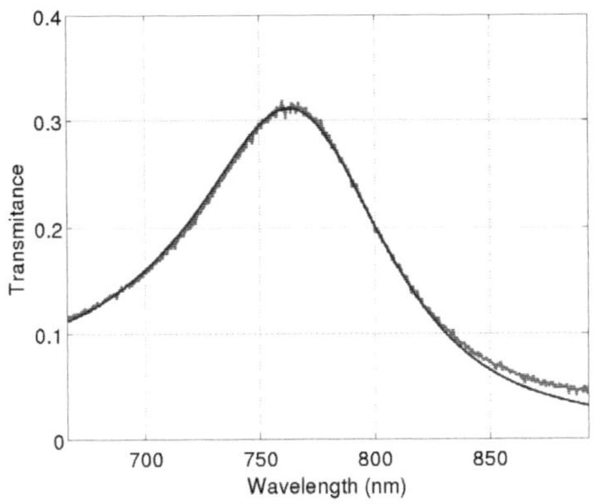

Figure 6: Fitting of theoretical and experimental transmittance of a transducer with ~22nm gold layer thickness. The matching ITO layer thickness is 144nm. The transmittance for λ_{probe} and λ_{pump} are 0.3 and less than 0.1 respectively. The experimental measurement is represented by a red curve while the theoretical calculation by a blue curve.

3. Experimental Results

3.1. Experimental set-up

Figure 5 shows a simplified schematic of the experimental set-up. The experiment is based on a dual laser ASOPS system [13]. The lasers consist on two 780nm, 150fs pulsed lasers with a repetition rate of 100MHz (10ns measuring window). This kind of system allows the tuning of the repetition rate of one of the lasers (probe) while locking its phase with respect to the other laser (pump). This allows the delay between the pulses to be set and swept without the need of a mechanical delay line. This provides a great simplification of the experimental set-up while increasing the acquisition speed. By having a 10kHz difference in repetition rate between the lasers, one full trace is acquired every 100μs. This means that it is possible to acquire up to 10k traces per second if the acquisition system is fast enough.

The pump laser is frequency doubled in a non linear crystal to 390nm. Both beams are combined by the use of a dichroic beamsplitter and focused on to the sample by a 50x long working distance objective. The probe beam is captured below the sample by a 20x long working distance objective and redirected to the detector by a dichroic beamsplitter that allows illumination for optical imaging on a CCD. The system typically uses 1mW average power for λ_{probe} and 0.5mW for λ_{pump} which corresponds to pulse energies of 10pJ and 5pJ respectively.

3.2. Brillouin scattering detection in transmission

The transducers were constructed by sputter coating the layers and their transmission was measured to compare it with a modelled one. As seen in figure 6, there is a good fit between theoretical and experimental results. ~0.3 transmission was reached for a device with ~20-140-20nm of gold-ITO-gold. For this device a ~27% of the input light makes it to the detector. This is almost three times more light compared to the ~0.1 of light returning on the reflection path in our system. This reduction in losses improves the SNR for the same amount of probe power coming out of the objective. Moreover, if the same probe power is going through the specimen by compensating the power coming out of the objective, the detection levels are significantly better for the transmission case since the losses in the path to the detector are smaller.

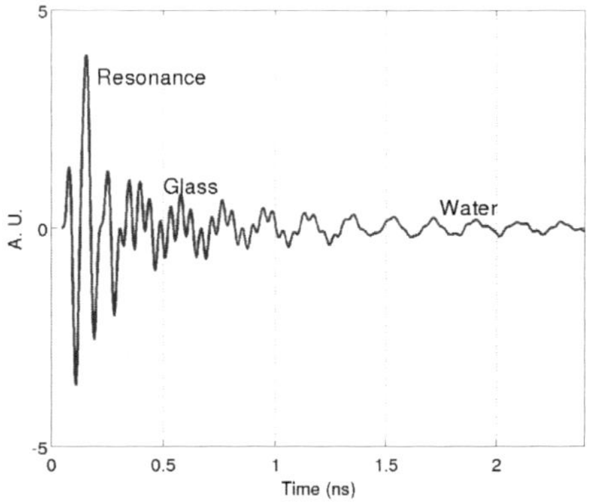

(a)

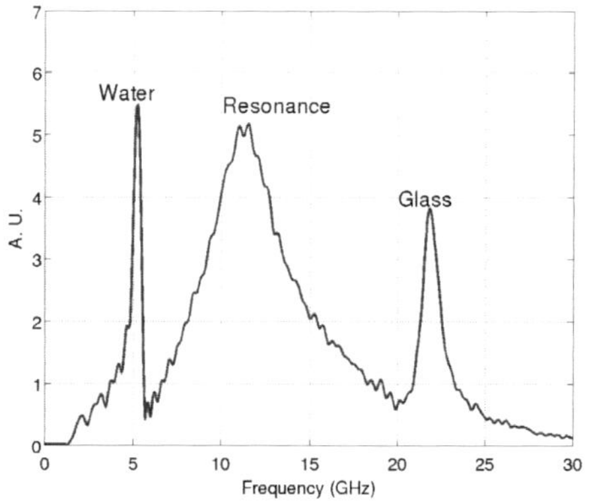

(b)

Figure 7: Sample of experimentally detected time traces from a water film in transmission using thin-film transducers. (a) Time trace where three signals are present; transducer response which decays quickly(~0.3ns) , the high frequency Brillouin oscillation from the glass substrate(~22GHz) and a low frequency Brillouin oscillation from the water film (~5GHz). (b) The Fourier transform showing the Brillouin peaks from the different signals.

A water film was used to test our approach due to its similarity to the cell media. The two Brillouin signals expected in these experiments from scattered light in the specimen and the glass substrate, are shown in figure 7. The mechanical resonance of the transducer is also detected due to the optical sensitivity of the transducer [2]. The time trace on figure 7a shows the change in instantaneous intensity of the probe beam as a result of the exposure to a pump pulse. There, the raw time trace was processed to remove the thermal background. On figure 7b, the Fourier transform of the trace in figure 7a is shown. The two sharp peaks correspond to the Brillouin oscillations in water (~5GHz) and glass (~22GHz). The additional broad peak is the device resonance which is due to the change in cavity size induced by the generated acoustic wave. The unwanted Brillouin component in the glass can be removed by digital filtering. The device resonance could be removed in a similar way or as it only last ~0.3ns, could be also removed by truncation.

3.3. Brillouin imaging of polystyrene phantoms

Before imaging a real cell, an experiment on phantom cells made out of melted lumps of 5μm polystyrene beads was performed. This provides a well known homogeneous specimen to test our approach. The beads were deposited on a transducer and then heated to 245 degrees centigrade for approximately 30 minutes to melt them. This leaves structures that resemble the shape of a

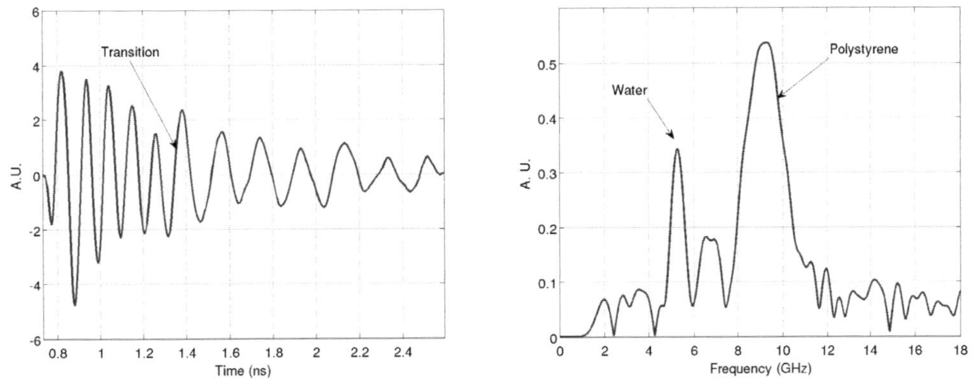

(a) (b)

Figure 8: Sample of experimentally detected traces in transmission using thin-film transducers on a polystyrene phantom cell. (a) Time trace where two frequencies are present; the high frequency Brillouin oscillation from the polystyrene phantom(~9GHz) and a low frequency Brillouin oscillation from the water surroundings (~5GHz). (b) The Fourier transform showing the spectral separation between the Brillouin peaks coming from the phantom and the medium.

cell and ensures good attachment to the transducer. The fabricated phantoms were submersed in water to simulate the presence of cell media that gives a background reference for contrast. The sample was mounted on a sealed chamber with the phantom facing the transmission objective (20x). Figure 8a shows an example of a time trace from a polystyrene phantom. There, it is possible to see the Brillouin frequency shifting from high frequency (9GHz) to low frequency (5GHz) when the sound leaves the phantom and enters the water. This transition is an indication of the phantom thickness at that particular point. The large difference in properties from water to polystyrene makes the transition easy to see however in the case of a real cell, the shift of the Brillouin Frequency will be much smaller.

Figure 9 shows an image of the Brillouin frequency of a polystyrene phantom. The scan consisted of 441 points covering 40 by 40μm in 2μm steps. The scan lasted approximately one hour taking 30k averages per point. A time trace was obtained at each point and its Brillouin frequency evaluated by performing a Fourier transform. Figure 9a shows an optical picture of the polystyrene phantom; the shape of the polystyrene beads is still visible. In figure 9b the Brillouin frequency map of the phantom shows good correlation with the optical picture. The variation of the Brillouin frequency within the phantom is small suggesting good homogeneity. If the refractive index in the phantom is constant (1.58 [14]), the speed of sound on the polystyrene obtained from equation 1 is approximately 2200 m/s, similar to the 2400m/s bulk speed reported by Smith et al [15].

3.4. Brillouin imaging of fixed cells

As demonstrated in the previous section, the approach presented in this work is viable. However, there are significant differences for a real cell; the Brillouin shift in the cell-media boundary is much smaller having the effect of reducing contrast. Also, the thickness of a cell is highly variable from sub micron in the edges to several tens of microns in the centre. These conditions make a cell a challenging subject.

3T3 fibroblast cells were selected for imaging due to their availability and their suitable dimensions. The transducers were pretreated with Poly-L-lysin, then the cells were seeded on

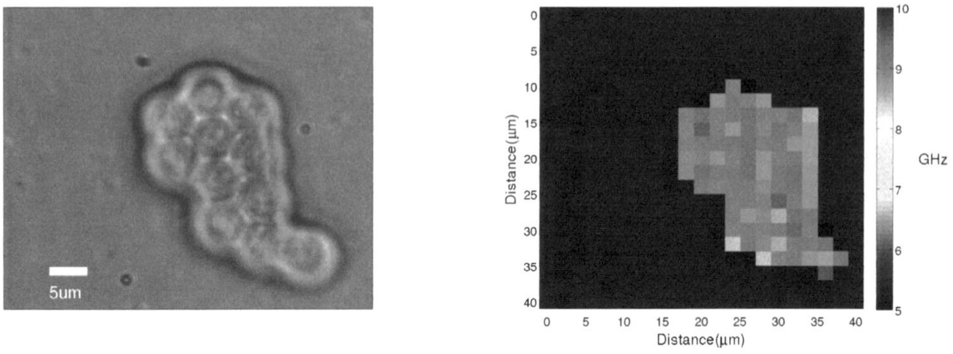

(a) (b)

Figure 9: Image of the Brillouin frequency of a phantom cell. (a) Optical picture of the phantom cell where the beads used to build it are partially visible. (b) The Brillouin map of the phantom cell. The correlation between optical and Brillouin frequency is good. There is large Brillouin frequency shift (~4GHz) due to the difference in material properties between polystyrene and water.

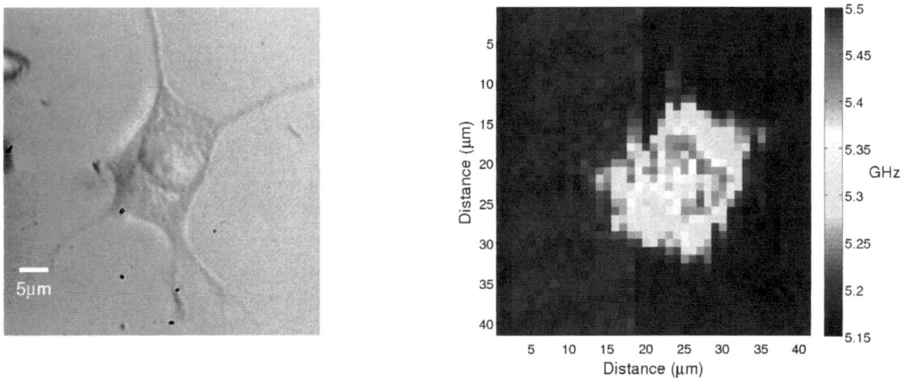

(a) (b)

Figure 10: Image of the Brillouin frequency of a fixed cell. (a) Optical picture of the fixed cell under examination. (b) The Brillouin map of the fixed cell. The correlation between the optical and Brillouin frequency image is good. The Brillouin frequency change in the cell is small (~20-250MHz) due to the similarity in mechanical properties between cell and medium.

them and fixed a day later. By assembling a two coverslip chamber (Live Cell Instruments Inc. chamilde CF-T) media was able to be pumped preventing bubble formation. Using the same scanning parameters and laser powers as the phantom case, an image of a fixed 3T3 cell was obtained. Figure 10 shows a fixed 3T3 fibroblast cell and its Brillouin frequency map. The correlation of the optical image with acoustical one is good. The features shown inside the cell in figure 10b are likely to be real changes in the properties of the cell. The measured Brillouin frequencies in the cell were between 5.25 and 5.5GHz with 50 and 250MHz shift from the Brillouin frequency of the background. Taking into consideration a refractive index of 1.36 [16] and assuming it is constant over the cell, the speed of sound in the cell changes from 1548 to 1605 m/s. Near the edges of the cell, the features are hard to see mostly because the signal lasts for a very short time and the signal from the media dominates the Fourier transform response.

4. Discussion

Since it is possible to observe when the sound waves leave the specimen and enter the water, it is also possible to calculate the thickness of the specimen if the speed of sound of is known. By knowing the refractive index before hand, it is easy to calculate the speed of sound using the Brillouin frequency. This opens the opportunity to not only measure the speed of sound but also to obtain the profile of the sample provided that the penetration depth of the acoustic wave is large enough to reach the boundary of a given object. It would be also possible to determine whether the cell is attached to the substrate or not.

Pulsed lasers emit continuously a train of pulses whether the acquisition system is ready or not. If the acquisition system is not ready to capture another set of traces when the trigger signal is received, then the signal is ignored and the specimen is exposed unnecessarily to laser light. The acquisition system could be busy processing data, sending it to a hard drive or simply waiting for the stages to move. Whatever the case our system can handle only approximately 30% of the pulses leaving the rest to be unnecessarily exposed to the specimen. There is an opportunity to improve this by blocking the pulses when the system is not ready and/or increasing the capabilities of the acquisition system and hence the capability of imaging a object which is changing in time.

The technique presented here could be extended for imaging living cells. However, the fragility of living cells introduces much stricter requirements on the light and heat doses delivered to the cells than those for fixed or phantom cells presented here. The thin film transducers presented in this paper go a long way to managing the light and heat load, but additional improvements can be made. With regards to light shielding, the current transducer blocks around 90% of the pump laser which allows images to be taken, but by changing the materials used for the dielectric layer (for example using SiO2), the transmission at the pump wavelength can be reduced (5%) giving greater protection for the cell . Heat exposure can also be reduced by the use of a different substrate with a higher thermal conductivity. For instance, sapphire offers 30 times higher thermal conductivity than a typical glass cover slip promoting better dissipation of heat and thus lower peak temperatures at the laser spot focus. These modifications will allow Brillouin oscillation images to be taken of living cells

4.1. Summary

In this paper images of the Brillouin frequency on phantom and fixed cells were presented by re-engineering thin film transducers [2]. The detection of Brillouin scattering in transmission was proposed and successfully applied to perform picosecond laser ultrasound experiments on delicate cultured cells. The approach is particularly useful for cells because the transducer itself act like a shield against pump laser light radiation while enhancing SNR. Brillouin frequencies measured on fixed 3T3 fibroblast cells were between 5.25 and 5.5GHz to calculate speed of sound from 1548 to 1605m/s which are similar to previously reported results on HeLa cells using a SAM (at 0.86GHz) [17] and on vegetal cells using Brillouin scattering detection [10].

References

[1] Kuznetsova et al 2007 *Micron* **137** 219
[2] Smith, Richard J. and Cota, Fernando Perez and Marques, Leonel and Chen, Xuesheng and Arca, Ahmet and Webb, Kevin and Aylott, Jonathon and Somekh, Micheal G. and Clark, Matt 2015 *JASA* **38** 824
[3] Nawaz et al 2012 *PloS one* **7** e45297
[4] Hochmuth, R M 2000 *J. Biomech* **33** 15
[5] Hildebrand, J A and Rugar, D and Johnston, R N and Quate, C F 1981 *Proc. Natl. Acad. Sci.* ***U S A*** **78** 1656
[6] Kundu, T and Bereiter-Hahn, J and Karl, I 2000 *Biophys. J.* **78** 2270

[7] Scruby, C.B and Drain, L.E 1990 *Laser Ultrasoncs Thecniques and Applications* (New York: Taylor and Francis Group) p 433
[8] Maznev, A A and Manke, Kara J and Lin, Kung-Hsuan and Nelson, Keith A and Sun, Chi-Kuang and Chyi, Jen-Inn 2012 *Ultrasonics* **52** 1
[9] Thomsen, C. and Grahn, H.T. and Maris, H.J. and Tauc, J. 1986 *Opt. Commun.* **60** 55
[10] Rossignol, C. and Chigarev, N. and Ducousso, M. and Audoin, B. and Forget, G. and Guillemot, F. and Durrieu, M. C. 2008 *Appl. Phys. Lett.* **93** 123901
[11] Ducousso, Mathieu and El-Farouk Zouani, Omar and Chanseau, Christel and Chollet, Céline and Rossignol, Clément and Audoin, Bertrand and Durrieu, Marie-Christine 2013 *EPJAP* **61** 11201
[12] Dehoux, T. and Audoin, B. 2012 *JAP* **112** 120702
[13] Elzinga, Paul A and Lytle, Fred E. and Jian, Yanan and King, Galen B. and Laurendeau, Normand M. 1987 *Appl. Spect.* **41** 2
[14] Xiaoyan Ma and Jun Q Lu and R Scott Brock and Kenneth M Jacobs and Ping Yang and Xin-Hua Hu 2003 *PMB* **48** 4165
[15] D. M. Smith and T. A. Wiggins 1972 *Appl. Opt.* **11** 2680
[16] Lanni, F and Waggoner, A S and Taylor, D L 1985 *J. Cell Biol.* **100** 219
[17] Weiss, Eike and Anastasiadis, Pavlos and Pilarczyk, Gotz and Lemor, Robert and Zinin, Pavel 2007 *IEEE UFFC* **54** 2257

An efficient model to predict guided wave radiation by finite-sized sources in multilayered anisotropic plates with account of caustics

M Stévenin[1,2], A Lhémery[1] and S Grondel[2]

[1] CEA LIST, point courrier 120, bâtiment 565, 91191 Gif-Sur-Yvette Cedex, France
[2] IEMN-DOAE, UVHC, Le Mont Houy, 59313 Valenciennes Cedex 9, France

E-mail: alain.lhemery@cea.fr

Abstract. Elastic guided waves (GW) are used in various non-destructive testing (NDT) methods to inspect plate-like structures, generated by finite-sized transducers. Thanks to GW long range propagation, using a few transducers at permanent positions can provide a full coverage of the plate. Transducer diffraction effects take place, leading to complex radiated fields. Optimizing transducers positioning makes it necessary to accurately predict the GW field radiated by a transducer. Fraunhofer-like approximations applied to GW in isotropic homogeneous plates lead to fast and accurate field computation but can fail when applied to multi-layered anisotropic composite plates, as shown by some examples given.
Here, a model is proposed for composite plates, based on the computation of the approximate Green's tensor describing modal propagation from a source point, with account of caustics typically seen when strong anisotropy is concerned. Modal solutions are otherwise obtained by the Semi-Analytic Finite Element method. Transducer diffraction effects are accounted for by means of an angular integration over the transducer surface as seen from the calculation point, that is, over energy paths involved, which are mode-dependent. The model is validated by comparing its predictions with those computed by means of a full convolution integration of the Green's tensor with the source over transducer surface. Examples given concern disk and rectangular shaped transducers commonly used in NDT.

1. Introduction

In most non-destructive testing (NDT) methods used to inspect plate-like structures by means of elastic guided waves (GW), transducers of finite size are used to generate these waves in the structure. GW propagate at long range, so that inspecting a large structure may be made by using a few permanently attached transducers, avoiding any scan to cover the whole structure. However, transducer diffraction effects take place, leading to complex field radiation characteristics. The optimization of transducer positioning for insuring the expected coverage must be carefully addressed and makes it necessary to accurately predict the guided wave field radiated by a transducer. In the literature, Fraunhofer-like approximations [1] have been developed for elastic guided waves in the case of isotropic and homogeneous plates, leading to fast computation of typical diffraction effects. In the case of multi-layered anisotropic plates such as those used in the aircraft industry, this approach fails at predicting the field radiated; some examples are given herein to illustrate this point. To solve the problem of efficient computing of transducer diffraction effect in GW radiation into multilayered anisotropic plates, a new model is proposed. It is based on the computation of the approximate Green's tensor describing modal propagation from a source point, with account of caustics which arise in anisotropic plates. The modal solution itself is obtained thanks to well-established Semi-Analytic Finite Element method [2]. The overall principle of the method is to proceed to an angular integration over the transducer surface as seen from the calculation point, based upon the energy paths involved, which are mode-dependent. The first part of the paper is dedicated to a description of the proposed model. In the second part, the model is validated thanks to comparisons of its predictions to those computed by means of a full convolution integral of the Green's tensor with the source description. Examples given concern disk and rectangular shaped transducers which are the most commonly used sources in the NDT practice.

2. Formulation of the 3D Green's function

2.1. Geometry of the system

We consider an infinite plate as shown in Fig. 1, of thickness $2d$, which upper plane is at $z = d$ and lower plane at $z = -d$. The plate is assumed to be made of an arbitrary composite material described as multiple parallel layers of arbitrary anisotropy.

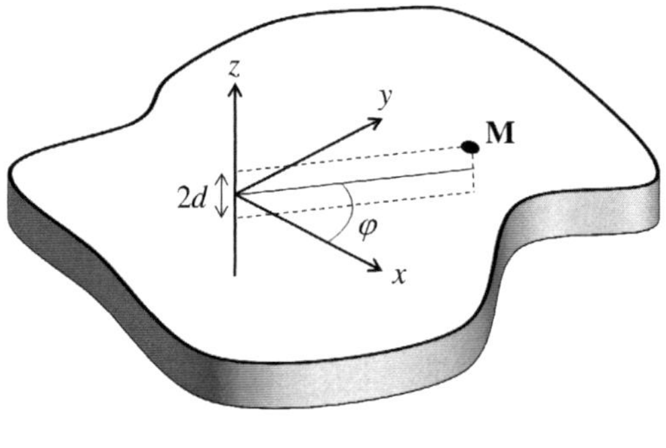

Figure 1. System geometry.

2.2. Far field expression

The field radiated by a finite-sized source can be expressed as a convolution integral of the Green's function $g^{(3)}(x, y, z)$ with a source term $q^{(3)}(x, y)$:

$$u^{(3)}(x, y, z, q^{(3)}) = \iint_S g^{(3)}(x - x', y - y', z) q^{(3)}(x', y') dx'dy'. \tag{1}$$

This Green's function can be written as a sum over the propagating modes [3]

$$g^{(3)}(x, y, z) = \sum_m \sum_{n(\varphi)} g^{(3)}_{m,n}(x, y, z). \tag{2}$$

These modal contributions are calculated thanks to the Semi Analytic Finite Element (SAFE) method [2]. With this time harmonic method, arbitrary sequences of multi-layered composite plates can be accounted for, results being given as propagative and evanescent modes described by their wavenumber and modal amplitude of components of the displacement along the plate thickness, for one given phase direction. The calculation is made once for all possible phase directions and results can be easily reused. The n subscript is never necessary when using isotropic plates. But, in some anisotropic cases it appears in several observation directions φ that for certain modes there are more than one phase contribution. Caustics can also appear for these modes.

To illustrate this, Fig. 2 presents the group velocity in $mm.\mu s^{-1}$ of the SH0 mode at a frequency of 300 kHz in a plate whose characteristics are given on the section 3.2. For this mode at this frequency we must deal with caustics. On the left, in the considered direction of observation, there are three phase contributions. On the right, in the considered direction of observation, two phase contributions coalesce, leading to a caustic.

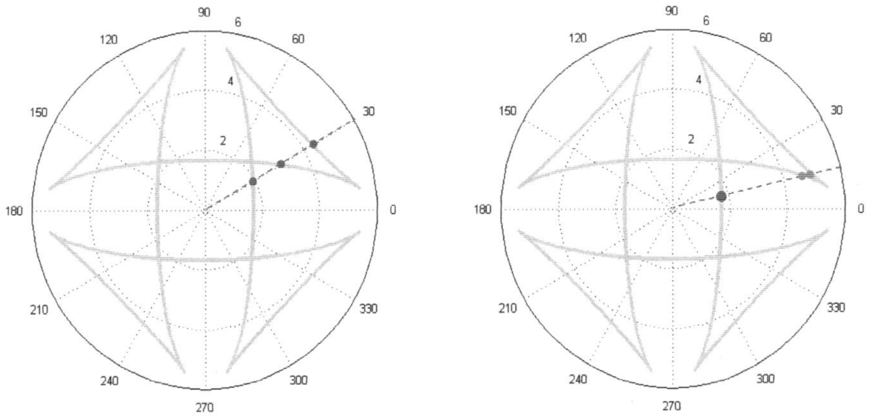

Figure 2. Three phase contributions far from caustics (left) and near a caustic (right).

To calculate the Green's function, one must distinguish propagation directions close to a caustic from the others. Far from caustic the phase term can be developed at the second order approximation whereas near a caustic, a third order developed must be used. The expressions of these Green's functions can be found in literature [3, 4] and are the following:

$$g^{(3)}_{m,n}(x,y,z) = d(x,y,z)a_{m,n}(x,y,z)p_{m,n}(x,y,z), \tag{3}$$

with,

$$d_f(x,y) = \left(\sqrt{x^2+y^2}\right)^{-1/2}, \tag{4}$$

$$a_{fm,n}(x,y,z) = res[G]_{k=k_{m,n}(\gamma_{m,n})} \frac{|k_{m,n}(\gamma_{m,n})|}{\sqrt{2\pi\left|\frac{\partial^2\phi_{m,n}(\gamma_{m,n},\varphi)}{\partial\gamma^2}\right|}}, \tag{5}$$

where f index stands for "far from caustics" and with G being 2D spatial Fourier transform of g,
$\phi_{m,n}(\gamma_{m,n},\varphi) = k_{m,n}(\gamma_{m,n})cos(\gamma_{m,n}-\varphi)$,
$k_{m,n}$ is the wave number of the mode,
and $\gamma_{m,n}$ is the phase direction for a given direction of observation φ, and where p denotes the following phase term

$$p_{fm,n}(x,y,z) = e^{i\sqrt{x^2+y^2}\phi_{m,n}(\gamma_{m,n},\varphi)} e^{i\frac{\pi}{4}sgn\left(\frac{\partial^2\phi_{m,n}(\gamma_{m,n},\varphi)}{\partial\gamma^2}\right)} e^{i\frac{\pi}{2}}, \tag{6}$$

far from caustics as presented on Fig. 2 and

$$d_c(x,y,z) = \left(\sqrt{x^2+y^2}\right)^{-1/3} \tag{7}$$

$$\begin{aligned} a_{cm,(1)+(2)}(x,y,z) &= res[G]_{k=k_{m,1}(\gamma_{m,1})}|k_{m,1}(\gamma_{m,1})|\left|\sqrt{\frac{-2\sqrt{S(\gamma_{m,1},\varphi)}}{\frac{\partial^2\phi_{m,1}(\gamma_{m,1},\varphi)}{\partial\gamma^2}}}\right| \\ &+ res[G]_{k=k_{m,2}(\gamma_{m,2})}|k_{m,2}(\gamma_{m,2})|\left|\sqrt{\frac{-2\sqrt{S(\gamma_{m,2},\varphi)}}{\frac{\partial^2\phi_{m,2}(\gamma_{m,2},\varphi)}{\partial\gamma^2}}}\right| \end{aligned} \tag{8}$$

$$p_{cm,(1)+(2)}(x,y,z) = e^{i\sqrt{x^2+y^2}L(\varphi)}Ai(\kappa)\frac{e^{i\frac{\pi}{2}}}{2} \tag{9}$$

near caustics when the two phase contributions (1) and (2) cannot be distinguished and treated separately as interfering contributions, as seen on Fig. 2, with

$L(\varphi) = \frac{1}{2}\left(\phi_{m,1}(\gamma_{m,1},\varphi) + \phi_{m,2}(\gamma_{m,2},\varphi)\right),$

$\kappa = -(\sqrt{x^2+y^2})^{\frac{2}{3}}S(\varphi),$

$S(\varphi) = \left[\frac{3}{4}\left(\phi_{m,1}(\gamma_{m,1},\varphi) + \phi_{m,2}(\gamma_{m,2},\varphi)\right)\right]^{\frac{2}{3}},$

Ai the Airy's function and where c index stands for "close to caustics".

Finally, we want to calculate the following integral:

$$u^{(3)}(x,y,z,q^{(3)}) = \sum_m \sum_{n(\varphi)} \iint_S g^{(3)}_{m,n}(x-x',y-y',z)q^{(3)}(x',y')dx'dy'. \tag{10}$$

In the isotropic case, such an integral can be accurately and efficently evaluated thanks to the Fraunhofer approximation [1]. In the next section, a Fraunhofer-like approximation is proposed for the anisotropic case.

3. Fraunhofer-like approximation

3.1. Theory

To start with this approximation, modes having only one phase contribution in each direction are considered. The source is centred on (x_c, y_c) and radiates with a uniform amplitude, so that for all source points, one has:

$$q^{(3)}(x',y') = q^{(3)}(x_c,y_c). \tag{11}$$

The Fraunhofer approximation consists in approximating the phase term by its first order Taylor expansion and the amplitude term by its zero order Taylor expansion. By doing so, Eq. (3) can be re-written as:

$$\begin{aligned} &d_f(x-x',y-y')a_{fm,n}(x-x',y-y',z)q^{(3)}(x',y') \\ &= d_f(x-x_c,y-y_c)a_{fm,n}(x-x_c,y-y_c,z)q^{(3)}(x_c,y_c) \\ &= (\sqrt{(x-x_c)^2+(y-y_c)^2})^{-1/2} \times res[G]_{k=k_{m,n}(\gamma_{m,n})}\frac{|k_{m,n}(\gamma_{m,n})|}{\sqrt{2\pi\left|\frac{\partial^2\phi_{m,n}(\gamma_{m,n},\varphi)}{\partial\gamma^2}\right|}}q^{(3)}(x_c,y_c), \end{aligned} \tag{12}$$

$$p_{fm,n}(x-x',y-y',z) = e^{i\phi_{m,n}(\gamma_{m,n},\varphi_c)\sqrt{(x-x_c)^2+(y-y_c)^2}} e^{i\frac{\pi}{4}sgn(\frac{\partial^2\phi_{m,n}(\gamma_{m,n},\varphi_c)}{\partial\gamma^2})} e^{i\frac{\pi}{2}} \times e^{i\phi_{m,n}(\gamma_{m,n},\varphi_c)(\frac{(x-x_c)(x-x')}{\sqrt{(x-x_c)^2+(y-y_c)^2}}+\frac{(y-y_c)(y-y')}{\sqrt{(x-x_c)^2+(y-y_c)^2}})}. \tag{13}$$

Fields radiated by most commonly used disk or rectangular transducers are now considered.

Disk-shaped transducers
The transducer is of radius a. To ease the calculation, a polar coordinate system is now used, so that the radiation integral is now given by:

$$u^{(3)}(x,y,z,q^{(3)}) = \sum_m d_f(x-x_c,y-y_c)a_{fm,n}(x-x_c,y-y_c,z)q^{(3)}(x_c,y_c)e^{i\frac{\pi}{2}} \times e^{i\frac{\pi}{4}sgn(\frac{\partial^2\phi_{m,n}(\gamma_{m,n},\varphi_c)}{\partial\gamma^2})} \int_{-\pi}^{\pi}\int_0^a e^{i\phi_{m,n}(\gamma_{m,n},\varphi_c)R(1-cos(\theta-\varphi_c)\frac{r'}{R})} r'dr'd\theta, \tag{14}$$

where $R=\sqrt{(x-x_c)^2+(y-y_c)^2}$. Classical wave diffraction calculations lead to the following final expression:

$$u^{(3)}(x,y,z,q^{(3)}) = \sum_m d_f(x-x_c,y-y_c)a_{fm,n}(x-x_c,y-y_c,z)q^{(3)}(x_c,y_c)e^{i\frac{\pi}{2}} \times e^{i\frac{\pi}{4}sgn(\frac{\partial^2\phi_{m,n}(\gamma_{m,n},\varphi_c)}{\partial\gamma^2})} \pi a^2 \frac{2J_1(\phi_{m,n}(\gamma_{m,n},\varphi_c)a)}{\phi_{m,n}(\gamma_{m,n},\varphi_c)a} e^{i\phi_{m,n}(\gamma_{m,n},\varphi_c)R}, \tag{15}$$

where the amplitude is proportional to the Fourier transform of the circular aperture.

Rectangular-shaped transducers
A rectangular transducer of aperture $l \times L$ is considered, so that radiation integral to calculate is given by:

$$u^{(3)}(x,y,z,q^{(3)}) = \sum_m d_f(x-x_c,y-y_c)a_{fm,n}(x-x_c,y-y_c,z)q^{(3)}(x_c,y_c)e^{i\frac{\pi}{2}} \times e^{i\frac{\pi}{4}sgn(\frac{\partial^2\phi_{m,n}(\gamma_{m,n},\varphi_c)}{\partial\gamma^2})} e^{i\phi_{m,n}(\gamma_{m,n},\varphi)\sqrt{(x-x_c)^2+(y-y_c)^2}} \times \int_{x_c-l/2}^{x_c+l/2}\int_{y_c-L/2}^{y_c+L/2} e^{i\phi_{m,n}(\gamma_{m,n},\varphi)(\frac{(x-x_c)(x-x')}{\sqrt{(x-x_c)^2+(y-y_c)^2}}+\frac{(y-y_c)(y-y')}{\sqrt{(x-x_c)^2+(y-y_c)^2}})} dx'dy'. \tag{16}$$

Similarly, classical diffraction calculations lead to the following result:

$$u^{(3)}(x,y,z,q^{(3)}) = \sum_m d_f(x-x_c,y-y_c)a_{fm,n}(x-x_c,y-y_c,z)q^{(3)}(x_c,y_c)e^{i\frac{\pi}{2}} \times e^{i\frac{\pi}{4}sgn(\frac{\partial^2\phi_{m,n}(\gamma_{m,n},\varphi_c)}{\partial\gamma^2})} lLsinc(\frac{\phi_{m,n}(\gamma_{m,n},\varphi_c)(x-x_c)}{\sqrt{(x-x_c)^2+(y-y_c)^2}}\frac{l}{2}) \times sinc(\frac{\phi_{m,n}(\gamma_{m,n},\varphi_c)(y-y_c)}{\sqrt{(x-x_c)^2+(y-y_c)^2}}\frac{L}{2}) e^{i\phi_{m,n}(\gamma_{m,n},\varphi_c)\sqrt{(x-x_c)^2+(y-y_c)^2}}. \tag{17}$$

where the amplitude is now proportional to the Fourier transform of the rectangular aperture. If an isotropic plate is considered momentarily, above results simplify and are expressed as given by A. Raghavan and C.E.S Cesnik [1]. In this paper, the Fraunhofer approximation is shown to give very accurate results.

Table 1. Plate characteristics.

C11 (GPa)	C22=C33 (GPa)	C12=C13 (GPa)	C23 (GPa)	C44 (GPa)	C55=C66 (GPa)	Mass density (kg/m3)	Ply thickness (mm)
123.4	11.5	5.6	6.4	2.6	4.5	$1.6x10^3$	0.25

3.2. Simulations

Simulations made using above given results are now presented for a composite plate.

Plate characteristics

The plate is 1mm-thick, made of four parallel layers of fiber-reinforced polymer(CFRP) with the following symmetry $[0°/90°]S$(T700GC/M21 cross-ply composite). Material parameters are listed in Table 1.

Figure 3 presents the slowness curves of the three propagating modes which exist at our work frequency of 300 kHz, below the first cut-off frequency of the plate. For each of these modes the dependency of phase directions on the observation (energy) direction is plotted on Fig. 4.

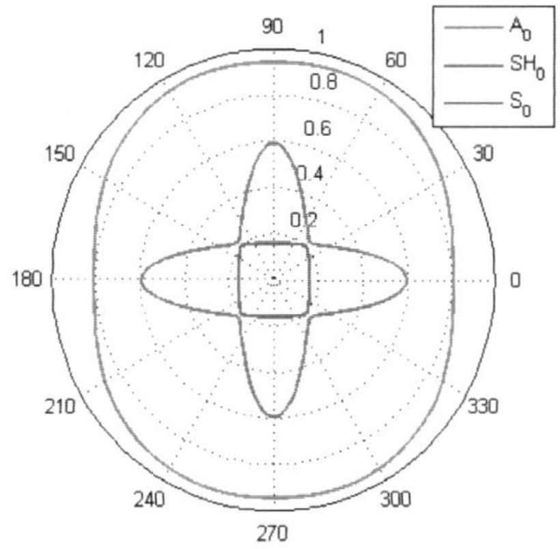

Figure 3. Slowness curves in $s.km^{-1}$ in a cross-ply $[0°/90°]S$ T700GC/M21 CFRP at a frequency of 300kHz.

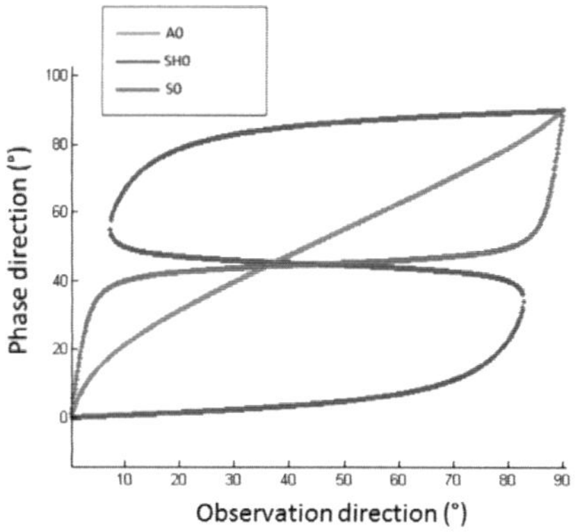

Figure 4. Phase directions function of the observation direction at a frequency of 300kHz.

Limits of the Fraunhofer like approximation

Results obtained in the previous section are now applied to the composite plate. In the slowness curves of Fig. 3, and considering the variations of phase directions with energy direction shown in Fig. 4, it appears that the A0 mode is the less anisotropic one: its slowness curve is quite circular and the direction of phase velocity is almost equal to that of the energy velocity. Therefore, the Fraunhofer approximation is expected to lead to accurate field prediction as it does in the isotropic case. In the computation, a 5-mm-diam disk transducer producing a uniform normal stress at the plate upper surface is considered. The computation is made for observation points located at 100mm of the source centre. Results computed using the approximate formula are compared to results obtained by computing the full surface convolution integral over the source active area.

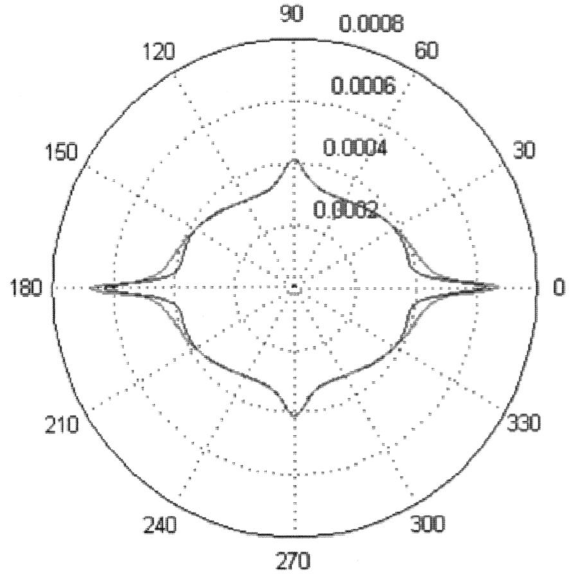

Figure 5. Out of plane displacement of the A0 mode at a distance of 100mm in the case of an excitation by a circular source at the frequency of 300 kHz obtained by computing the full integration (red line) and the approximate formula (blue line).

The two kinds of results superimpose quite well but there are angular regions where approximate results are clearly wrong. Discrepencies appear in directions where small variation of the energy direction (e.g 5°) leads to large variation of phase direction (15°). The sole phase direction used in the approximation does not permit to accurately account for phase variations over the source surface. As the mode considered is the less anisotropic, such a failure is of course even worse when more anisotropic modes are considered. It is therefore obvious that Fraunhofer approximation cannot be reliably used for anisotropic plates. To overcome this difficulty, still aiming at deriving computationally efficient formulae, a new model is now proposed.

4. Integration along energy directions

4.1. Theory

We will use a new coordinate system in this part, as shown by Fig. 6. The system is centred on the calculation point and rotated so that $\vec{e_x}$ is along the direction linking the calculation point to the source centre.

In the new coordinate system, the source centre is at $(R, 0)$ with $R = \sqrt{(x - x_c)^2 + (y - y_c)^2}$. The radiated field is now expressed as:

$$u^{(3)}(e_x = 0, e_y = 0, z, q^{(3)}) = \sum_m \sum_{n(\varphi)} \iint_S d(-e'_x, -e'_y, z) a_{m,n}(-e'_x, -e'_y, z) \times p_{m,n}(-e'_x, -e'_y, z) q^{(3)}(e'_x, e'_y) de'_x de'_y, \tag{18}$$

The field expression is reformulated to explicit the angular dependence on θ, angle formed by the line linking the calculation point to the source centre and that linking the calculation point to the sourcepoint:

$$\theta = \varphi' - \varphi_c, \tag{19}$$

with φ' the direction between the source point and the observation point and φ_c direction between the centre of the source and the observation point in the former coordinate system.

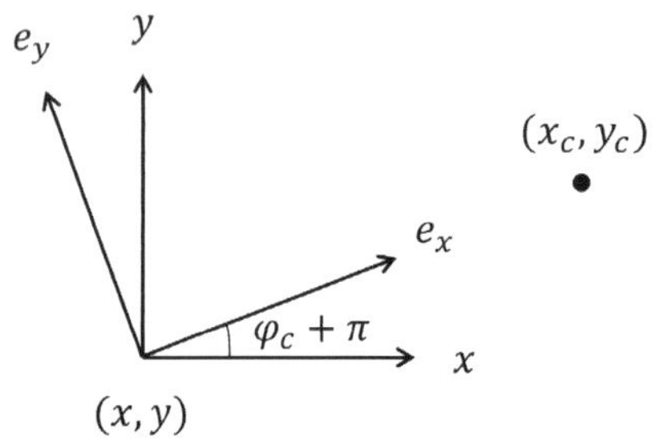

Figure 6. New coordinate system.

In our integral r is the distance between the source and calculation points. In this new formulation $d(x, y) = d(r)$:

$$\begin{aligned} u^{(3)}(e_x = 0, e_y = 0, z, q^{(3)}) = \sum_m \sum_{n(\varphi)} \int_{\theta_{min}}^{\theta_{max}} d\left(\frac{r_1(\theta) + r_2(\theta)}{2}\right) \\ \times \int_{r_1(\theta)}^{r_2(\theta)} a_{m,n}(\theta, z) p_{m,n}(r', \theta, z) q^{(3)}(r', \theta) r' dr' d\theta, \end{aligned} \tag{20}$$

$[\theta_{min}, \theta_{max}]$ and $[r_1(\theta), r_2(\theta)]$ denote the integration intervals as shown on the Fig. 7.

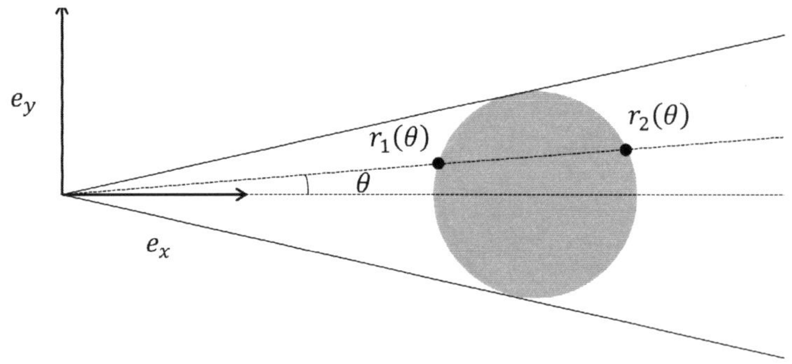

Figure 7. Integration limits.

Calculation points must be outside the source. Moreover, we still consider a uniform source distribution.

$$
\begin{aligned}
I_{m,n}(e_x=0,e_y=0) &= \int_{\theta_{min}}^{\theta_{max}} d\left(\frac{r_1(\theta)+r_2(\theta)}{2}\right)\int_{r_1(\theta)}^{r_2(\theta)} a_{m,n}(\theta,z)p_{m,n}(r',\theta,z)r'dr'd\theta \\
&= \sum_{t=1}^{n_\theta} \frac{\theta_{max}-\theta_{min}}{n_\theta} d\left(\frac{r_1(\theta_t)+r_2(\theta_t)}{2}\right) \\
&\times \int_{r_1(\theta_t)}^{r_2(\theta_t)} a_{m,n}(\theta_t,z)p_{m,n}(r',\theta_t,z)r'dr',
\end{aligned} \tag{21}
$$

with $\theta_t = \theta_{min} + (t-\frac{1}{2})\frac{\theta_{max}-\theta_{min}}{n_\theta}$ and n_θ the number of discrete directions in angular integration.

For directions far from a caustic, we have:

$$
\begin{aligned}
d&\left(\frac{r_1(\theta_t)+r_2(\theta_t)}{2}\right)\int_{r_1(\theta_t)}^{r_2(\theta_t)} a_{m,n}(\theta_t,z)p_{m,n}(r',\theta_t,z)r'dr' \\
&=d_f\left(\frac{r_1(\theta_t)+r_2(\theta_t)}{2}\right) e^{i\frac{\pi}{4}sgn(\frac{\partial^2\phi_{m,n}(\gamma_{m,n},\theta_t)}{\partial\gamma^2})} e^{i\frac{\pi}{2}} a_{fm,n}(\theta_t,z) \\
&\times\left(e^{ir_2(\theta_t)\phi_{m,n}(\gamma_{m,n},\theta_t)}\frac{ir_2(\theta_t)\phi_{m,n}(\gamma_{m,n},\theta_t)-1}{(i\phi_{m,n}(\gamma_{m,n},\theta_t))^2}\right. \\
&\left. -e^{ir_1(\theta_t)\phi_{m,n}(\gamma_{m,n},\theta_t)}\frac{ir_1(\theta_t)\phi_{m,n}(\gamma_{m,n},\theta_t)-1}{(i\phi_{m,n}(\gamma_{m,n},\theta_t))^2}\right).
\end{aligned} \tag{22}
$$

For directions close to a caustic, we have:

$$
\begin{aligned}
d&\left(\frac{r_1(\theta_t)+r_2(\theta_t)}{2}\right)\int_{r_1(\theta_t)}^{r_2(\theta_t)} a_{m,n}(\theta_t,z)p_{m,n}(r',\theta_t,z)r'dr' \\
&= d_c\left(\frac{r_1(\theta_t)+r_2(\theta_t)}{2}\right)\frac{e^{i\frac{\pi}{2}}}{2}Ai(\kappa(\theta_t))a_{cm,n}(\theta_t,z) \\
&\times\left(e^{ir_2(\theta_t)L(\theta_t)}\frac{ir_2(\theta_t)L(\theta_t)-1}{(iL(\theta_t))^2} - e^{ir_1(\theta_t)L(\theta_t)}\frac{ir_1(\theta_t)L(\theta_t)-1}{(iL(\theta_t))^2}\right),
\end{aligned} \tag{23}
$$

with, $\kappa(\theta_t) = -\left(\frac{r_1(\theta_t)+r_2(\theta_t)}{2}\right)^{\frac{2}{3}} S(\theta_t)$.

This general formulation involving integration along energy directions to express the field is now used for circular or rectangular sources. Integration intervals in calculation point coordinate system are easily found for both of them.

4.2. Results
To validate the proposed model of integration, results predicted by means of this model are compared to equivalent results computed by means of the full convolution surface integral. The cases of a disk transducer and of a rectangular one are considered. The plate considered is the same as that used in previous computations.

Disk-shaped transducers
A disk transducer of radius 5 mm is considered. Calculation points are located at a distance of 100mm from the source centre. The amplitude displayed is the absolute value of the out-of-plane component of the displacement.

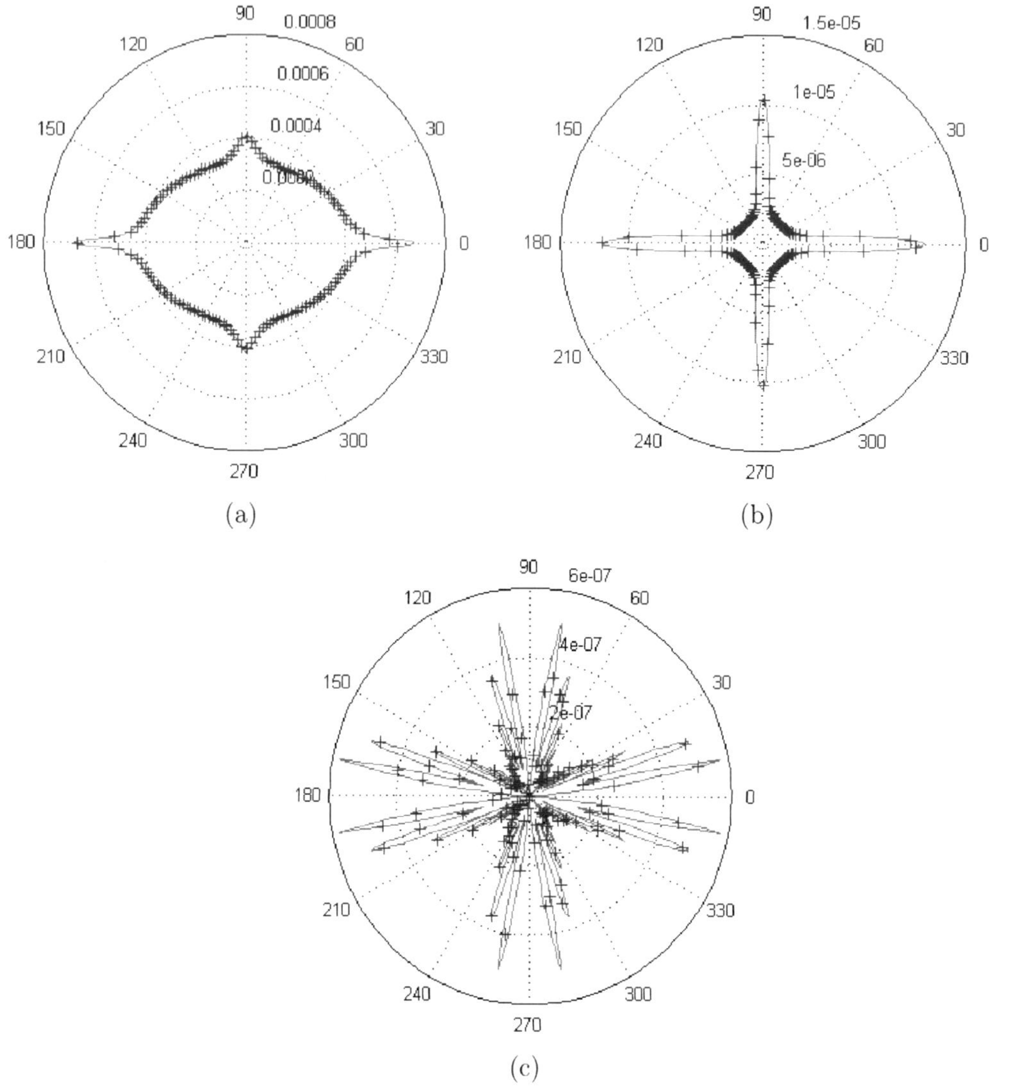

Figure 8. Out of plane displacement of the A0 mode (a), S0 mode (b), SH0 mode (c) at a distance of 100mm in the case of an excitation by a circular source at the frequency of 300 kHz obtained by computing the full integration (solid line) and the approximate formula (+ points).

Figure 8 (a) displays the same results as those shown on Fig. 5. In the previous results, the Fraunhofer approximation was shown to fail in some angular regions, despite the fact that the mode considered (A0) is the less anisotropic one among the three propagative modes existing below the first cut-off frequency. Here, the proposed model leads to predictions in excellent agreement with those computed using the full integration scheme. The two other comparisons shown on Fig. 8 (b, c) now concern the two other modes of higher anisotropy S0 and SH0. Even for the highest anisotropy (SH0), results predicted by the proposed model perfectly agree with those computed without approximation.

Rectangular-shaped transducers

Similar results are shown on Fig. 9 for a rectangular transducer (actually as square source of 9-mm-side). Directivities of the various modes are different from those radiated by the disk transducer. Again, results computed using the proposed model or using the full integration superimpose.

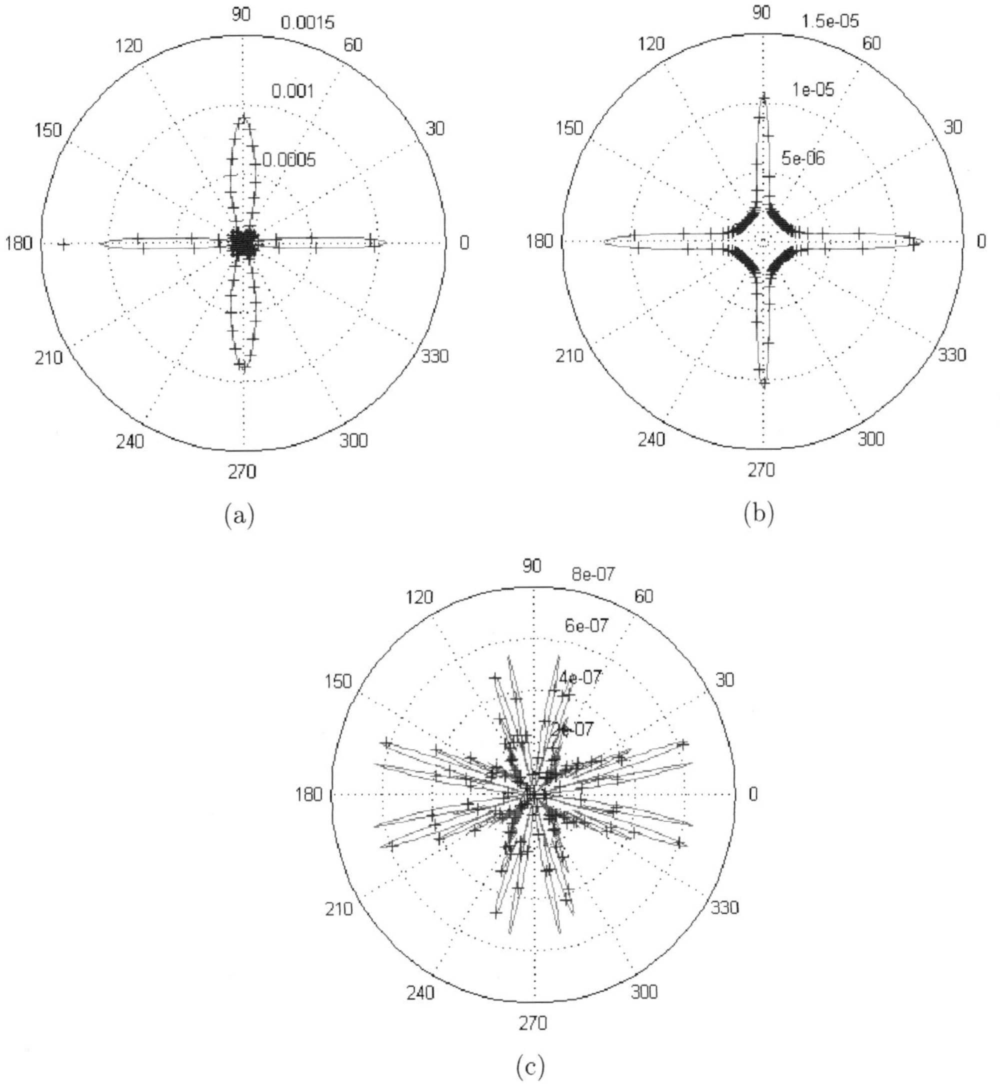

Figure 9. Out of plane displacement of the A0 mode (a), S0 mode (b), SH0 mode (c) at a distance of 100mm in the case of an excitation by a square source at the frequency of 300 kHz obtained by computing the full integration (solid line) and the approximate formula (+ points).

We can see on the Fig. 9 that the results, as for the circular shaped transducer are similar between the convolution and the integration along energy directions.

Note that in Fig. 8 and Fig. 9, results computed by means of the approximate formulation perfectly superimpose with those computed by means of the full integration, so much so that results shown as solid lines could not be distinguished. This is the reason for presenting the former results with "+" signs and to limit the number of angles for which they are shown.

5. Discussion

The proposed approximate model leads to very accurate prediction of results obtained by means of the full integral over the source aperture. The approximation is quite limited as it only concerns the amplitude term along the intersection of a ray with the source surface, the phase term being exactly taken into account and analytically integrated along this intersection. The angular integration over ray directions covers the full surface seen from the computation point. With the present model, a 1D numerical integration is involved, this leading of course to faster computation than that required for the full 2D one. Moreover, the farther the computation point, the smaller the angular range of integration and the faster the numerical evaluation of the integral.

6. Conclusion

Fields of guided waves radiated by finite-sized sources must be evaluated to optimize testing configurations of plate-like structures. Fraunhofer approximation is known to lead to very accurate predictions of transducer diffraction effect for a minimum computational effort, but this is only true if the plate is isotropic. As soon as the plate structure is anisotropic, all the guided modes generated by a source behave anisotropically. For a given multi-layered composite plate, the anisotropy varies a lot from one mode to another. Some results have been shown in the first part of the paper proving that, even for the less anisotropic mode, Fraunhofer approximation fails at predicting accurately the field radiated. This makes it necessary to account for all the possible energy direction. The existence of caustics is an issue that was recently solved in the literature in the form of an approximate Green's function. The model proposed in the paper uses the modal description computed by the SAFE method and expressions of the Green's function which differ if the energy direction is close to or far from a caustic [4]. A change of variables in the convolution surface integral expressing the field was proposed to optimally compute this integral (at low cost), leading to expressions easy to implement and fast computed. The model has been shown to predict very accurately the radiated field associated to various modes, even for the most anisotropic ones.

References

[1] Raghavan A and Cesnik C E S 2005 *Smart Mater. Struct.* **14** 1448
[2] Taupin L, Lhémery A and Inquité G 2011 *J. Phys. : Conf. Ser.* **269** 012002
[3] Velichko A and Wilcox P D 2007 *J. Acoust. Soc. Am.* **121** 60
[4] Karmazin A, Kirillova E, Seeman W and Syromyatnikov P 2013 *Ultrasonics* **53** 283

The dynamic excitation of a granular chain for biomedical ultrasound applications: contact mechanics finite element analysis and validation

P Gélat[1†], J. Yang[2], P J Thomas[2] D A Hutchins[2], O Akanji[2], L A J Davis[2], S Freear[3], S Harput[3] and N Saffari[1]

[1] UCL Mechanical Engineering, University College London, Torrington Place, London WC1E 7JE, United Kingdom.

[2]School of Engineering, University of Warwick, CV4 7AL, United Kingdom.

[3]School of Electronic and Electrical Engineering, University of Leeds, Leeds LS2 9JT, United Kingdom.

[†] E-mail: p.gelat@ucl.co.uk

Abstract. There has been recent interest in the transmission of acoustic signals along granular chains of spherical beads to produce waveforms of relevance to biomedical ultrasound applications. Hertzian contact between adjacent beads can introduce different harmonic content into the signal as it propagates. This transduction mechanism has the potential to be of use in both diagnostic and therapeutic ultrasound applications, and is the object of the study presented here. Although discrete dynamics models of this behaviour exist, a more comprehensive solution must be sought if changes in shape and deformation of individual beads are to be considered. Thus, the finite element method was used to investigate the dynamics of a granular chain of six, 1 mm diameter chrome steel spherical beads excited at one end using a sinusoidal displacement signal at 73 kHz. Output from this model was compared with the solution provided by the discrete dynamics model, and good overall agreement obtained. In addition, it was able to resolve the complex dynamics of the granular chain, including the multiple collisions which occur. It was demonstrated that under dynamic excitation conditions, the inability of discrete mechanics models to account for elastic deformation of the beads when these lose contact, could lead to discrepancies with experimental observations.

1. Background

Granular chains are usually defined as one-dimensional alignments of generally spherical particles. The interactions between such particles are governed by the laws of contact mechanics. The study of granular chains is of interest to a range of disciplines in science and engineering. Granular chains have been relevant to applications where they can serve as a type of waveguide and/or information carrier by transmitting, redirecting, or blocking solitary waves in a controllable and efficient manner [1]. Solitary waves haves been known to be generated in a one-dimensional granular chain of spherical particles, where dispersive and nonlinear effects, due to the discreteness of the system and the Hertzian contact among spheres balance out [2], [3], [4]. Through the use of a discrete mechanics model, solitary waves were first shown to be supported in a granular chain by Nesterenko in 1983 [2]. This theoretical observation was subsequently confirmed experimentally by Lazaridi and Nesterenko in 1985 [5]. Coste *et al* [6] carried out an experimental studied of the propagation of high-amplitude compressional waves in a chain of beads in Hertzian contact, submitted or not to a small pre-compressional static force. An extensive range of pulse amplitudes was investigated, and it was concluded that the shape and velocity the solitary waves as functions of their maximum amplitude were in good agreement with results from the discrete mechanics model.

Interest in the generation of solitary waves in a granular chain has recently been extended to biomedical applications [7], [8]. In [7], the generation of high-amplitude focused acoustic pulses using a one-dimensional array of granular chains was investigated, where the amplitude, size, and location of the focus could be controlled by varying the static pre-compression of the chains. Such an array could have important applications to both therapeutic high-intensity focused ultrasound applications and also perhaps in imaging. It is however yet unclear which advantages a transduction mechanism based on the non-linear dynamics of granular chains may have compared with more commonly used arrays of piezoelectric transducers, such as those described in [9]. Furthermore, the dimensions of the array as well as the frequencies involved in the configuration described in [7] are not in fact pertinent to biomedical ultrasound. Hence, further studies are clearly required, which involve excitation protocols which are relevant to clinical ultrasound applications.

In [8], large amplitude displacements were produced by a resonant 73 kHz ultrasonic source to drive a granular chain consisting of six 1 mm diameter chrome steel beads. Travelling solitary wave impulses were observed, which were due to both non-linearity between adjacent spheres and reflections within the chain. The resulting impulses in the axial velocity measured at the final bead of the chain were found to possess both high amplitude and wide bandwidth, featuring spectral content up to 200 kHz. These distinctive features suggest that such signals could be harnessed for both therapeutic and diagnostic ultrasound applications.

In [8], predictions of the bead dynamics based on the discrete mechanics model proposed by Lydon *et al* [10] reproduced characteristics of the laser vibrometer measurements of the velocity of the final bead of the granular chain. This observation, together with the comparisons in [5], [6] and [10] suggest that discrete mechanics models clearly replicate salient features of experimental observations. Nevertheless, to better understand and optimise the transduction process and make it application-specific, it is desirable to move beyond discrete mechanics formulations. Indeed, a transducer for biomedical ultrasound applications comprising a granular chain may include other components, such as piezoelectric actuator and matching layers. Furthermore, the transducer will couple into an acoustic medium, such as water or soft tissue, the loading of which will affect its dynamic behaviour. Hence, a numerical solution to the design of such a transducer is likely to provide more flexibility in designing an application-specific device.

Finite element analysis (FEA) is routinely used to solve contact mechanics problems in engineering [11]. Many commercially available FEA packages also allow for electro-mechanical coupling with piezoelectric elements, as well as fluid-structure coupling, and can therefore predict the radiated acoustic pressure for a known input voltage. Musson and Carlson [12] proposed an FEA of the non-linear dynamics of the granular chain investigated by Lazaradi and Nesterenko [5]. They demonstrated that solitary waves can be generated in a chain of 20 beads using FEA. This solution was provided by the COMSOL Multiphysics software [13], and the analysis demonstrated the importance of localised plastic deformations in the dynamics of granular chains. In discrete mechanics models, these plastic deformations are neglected, as the beads are assumed to be point masses [6].

The excitation of the first bead of the chain considered in [12] was of the same type as that was assumed in [5] and consisted of a single constant velocity impact. Whilst it is encouraging that the FEA generated improved agreement with the experimental work in [5] relative to the discrete mechanics model, further work is clearly required to extend this type of analysis to excitation signals of relevance to biomedical ultrasound. It is therefore of interest to investigate whether FEA can successfully predict the response of a granular chain to the high-amplitude and high-frequency

Journal of Physics: Conference Series **684** (2016) 012005 doi:10.1088/1742-6596/684/1/012005

excitation signals described in [8]. The work described in this paper constitutes a preliminary investigation where it was initially verified that Hertz' contact law [14] was satisfied for two spheres undergoing static compression. Subsequently, the six-bead granular chain described in [8] was excited with five cycles of a sinusoidal displacement, via a rigidly vibrating cylindrical piston. The amplitude of the displacement excitation was 0.3 μm. The frequency of excitation was also the same as the fundamental frequency of the excitation waveform used in [8], i.e. 73 kHz. The FEA was carried out using a transient analysis in ANSYS Mechanical version 15.0 [14]. Displacements at the centre of each bead were obtained as a function of time and compared with results from the discrete mechanics model described in [10], as implemented in [8], and in absence of any viscous damping. The FEA results were in good overall agreement with the discrete mechanics model. To investigate the source of the discrepancies between both models, the elastic deformations of the beads along the axis of the chain were analysed, as a function of time.

2. Theory

When the surfaces of two separate bodies touch each other so that they become mutually tangential, they are said to be in contact. In the physical sense, the surfaces that are in contact do not interpenetrate and can transmit compressive normal forces and tangential frictional forces. They do not generally transmit tensile normal forces and are thus free to separate and move away from each other.

The static frictionless interaction between two adjacent elastic spheres is an exact solution of linear elasticity and is known as Hertz's law [14]. As a result of geometrical effects, there exists a nonlinear relationship between the exerted force F_0 on the spheres and the distance of approach δ_0 of their centers. For homogeneous isotropic spheres of radius a, this is expressed as follows [6]:

$$\delta_0 = \frac{2(\theta F_0)^{2/3}}{a^{1/3}} \tag{1}$$

where

$$\theta = \frac{3(1-\upsilon^2)}{4E} \tag{2}$$

where E and υ are respectively Young's modulus and Poisson's ratio corresponding to sphere material.

In the case of a dynamic excitation, this solution is expected to remain valid when the applied force and hence the distance of approach are both slowly varying functions of time [6]. According to [6], the variations may be considered as slow if every typical time scale involved in the motion is much greater than the time needed by a bulk longitudinal acoustic wave to travel across the diameter of a bead. In such cases, a discrete mechanics model has been proposed to model a chain of N identical beads, which essentially consists of a one-dimensional system of point masses linked by nonlinear springs. This can be expressed as a system of coupled nonlinear differential equations, where the equation of motion of the ith bead is given by [6]:

$$\ddot{u}_i = \frac{1}{2m\theta}\sqrt{\frac{a}{2}}\left([\delta_0 - (u_i - u_{i-1})]^{3/2} - [\delta_0 - (u_{i+1} - u_i)]^{3/2}\right) \tag{3}$$

where u_i is the displacement at the centre of the ith bead and $\ddot{u}_i$ is its second derivative with respect to time. The mass m associated with each bead is equal to $\frac{4\pi\rho a^3}{3}$, where ρ is the density of the bead material. This formulation neglects dissipation effects which may arise from the beads' viscoelasticity. It also neglects the plastic deformation of the beads. An extension to this model was proposed by Lydon *et al* [10] to address the lack of dissipative effects in equation (3) and to better reflect potential experimental conditions. Indeed, although viscoelastic effects may be small in materials such as steel, the beads require a holder to keep them axially aligned during excitation. Such a holder is likely to introduce dissipative effects. To account for this, Lydon *et al* [10] introduce a linear viscous damping coefficient between successive spheres in the dynamic equations. The damping term is only relevant when the spheres are in contact and this is controlled by means of a Heaviside function. Loss of contacts between the beads can be accounted for by noting that only positive arguments of the 3⁄2 power-law terms in equation (3) need be considered. These terms can be set to zero for negative values of these arguments, i.e. tensionless behaviour, when the spheres lose contact. This model was used in [8] to simulate the dynamics of a chain of six chrome steel spheres in Hertzian contact. The model also considers contact of the first and last sphere of the chain with spheres of infinite radius, thus approximating walls of a semi-infinite domain. In the investigations carried out in [8] and [10], the first sphere is therefore effectively excited by a flat surface possessing the properties of the ultrasonic horn tip exciting the chain, as the last sphere is in contact with a static wall possessing the properties of the support. These contacts are assumed to be Hertzian in nature.

The work carried out by Hutchins *et al* [8] was used as a starting point for the dynamic analysis described in this paper. The granular chain consists of six spherical chrome steel beads. It was assumed that no pre-compressional static force was acting along the axis of the chain, so that the beads were just touching. Although this situation implies that no acoustic waves can propagate up and down the chain, it is known to lead to highly non-linear behaviour [6], which is more likely to generate signals relevant to biomedical ultrasound [7], [8]. All FEA was carried out using ANSYS™ Mechanical version 15. The beads were assumed to be perfectly aligned, so that an axisymmetric analysis could be opted for, whereby the Cartesian y-axis defined the axis of symmetry of the granular chain, and the analysis was carried out in the x-y plane. The first bead of the chain was excited via a cylindrical piston whose degrees of freedom were coupled to ensure rigid body motion. A rigidly vibrating piston was employed to reduce uncertainties when comparing with the discrete mechanics model in [8] and [10], since modelling semi-infinite structures using FEA presents challenges. For similar reasons, the final bead of the chain was assumed to be in contact with a rigid support. All contacts were assumed to be frictionless, thus simulating Hertzian contacts. The introduction of a linear viscous damping coefficient between successive spheres, as proposed in [10], was not implemented as part of this work. Again, including this type of damping in the FEA model would add a layer of complexity to the validation process. Hence, dissipative mechanisms will be neglected here and dealt with in a separate paper.

It is common to formulate the problem of frictionless contact between two solid bodies as a variational inequality. This presents a special type of minimisation problem with inequality constraints, which can be efficiently treated in a standard manner, i.e. with (1) the penalty method, (2) the augmented Lagrangian method or (3) the Lagrange multiplier method [11]. The ANSYS™ Mechanical v15.0 FEA package includes these three options as formulations to establish a relationship between two surfaces to prevent or limit them from passing through each other during the analysis. For an in-depth description of these methods, the reader is referred to Chapter 4 of [11]. Methods (1) and (2) are both penalty-based and invoke the concept of contact stiffness i.e. the virtual work due to the deformation of imaginary springs at the contact interface term. This inevitably results in some degree of penetration between the two surfaces, depending on the chosen value for the contact stiffness term. Method (3), or the Normal Lagrange Formulation as it is described in ANSYS Mechanical, adds an extra degree of freedom (contact pressure) to satisfy contact compatibility. Consequently, instead of resolving contact force as contact stiffness and penetration, contact pressure is solved for explicitly as an extra degree of freedom. This has the advantage of enforcing near-zero penetration when modelling frictionless contact between two bodies. This formulation however requires the use of a direct rather than an iterative solver, which may limit the size of the models solved due to larger RAM requirements. Fortunately, the geometry of granular chain could be defined using an axisymmetric coordinate system. Hence, computational requirements were not an impeding factor.

Given the complexity of the dynamics of the six-bead granular chain under the excitation conditions described in [8], it is likely that any effects of penetration between two adjacent surfaces will lead to uncertainties in the displacements of the beads that will propagate throughout the system as time increases. This will constitute a source of disagreement between the FEA and the discrete mechanics model. For this reason, the Normal Lagrange Formulation was opted for over the penalty and augmented Lagrangian formulations.

3. Results

Prior to carrying out a dynamic FEA of the six-bead granular chain described in Section 2, the case of the static compression of two 1 mm diameter chrome steel spheres was investigated so as to determine to which extent Hertz' contact law was satisfied. The configuration was assumed to be axisymmetric, the Cartesian y-axis being the axis of symmetry. The top-most sphere was stressed by a static force, applied in the negative y-direction, as shown in figure 1. The bottom-most sphere was assumed to be in Hertzian contact with a rigid support. The region of contact common to both spheres was also assumed to be Hertzian. The applied force was varied between 0.5 N and 5 N. The material properties of the sphere material were the same as those assumed by Hutchins *et al* (2015), i.e.:

- Young's modulus: 201 GPa;
- Density: 7833 kg m^{-3};
- Poisson's ratio: 0.3.

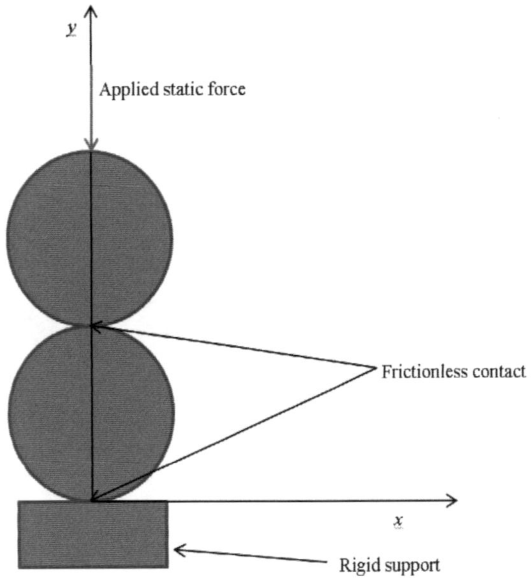

Figure 1. Schematic diagram of two-sphere configuration.

Particular attention was given to the mesh density around the contact regions. Indeed, contact mechanics problems in FEA tend to require mesh refinements around such regions [12], [15]. A close-up view of the mesh around the contact between both spheres is shown in figure 2.

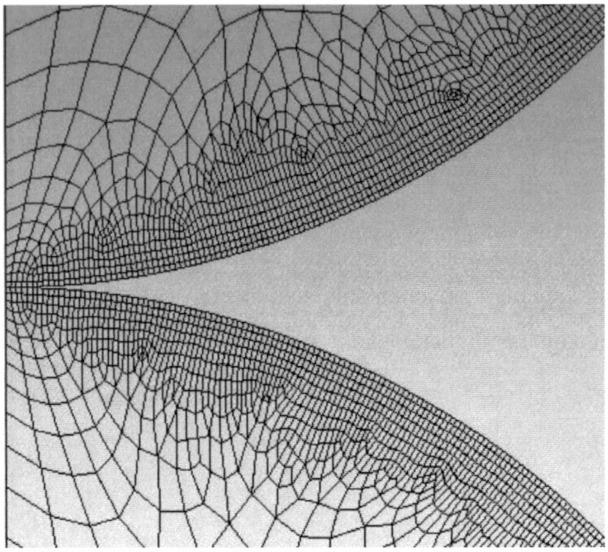

Figure 2. Close-up of mesh around contact region between both spheres.

When two spheres in Hertzian contact each undergo a static force F_0, the distance of approach between the centre of each sphere is given by equation (1) [6]. After post-processing the displacements at the centres of the spheres the *y*-direction, the distance of approach was calculated and compared with the corresponding analytical solution predicted by equation (1). The distance of approach as a function of the applied static force is shown in figure 3. Both the analytical solution obtained from equation (1), as well as the FEA results are shown.

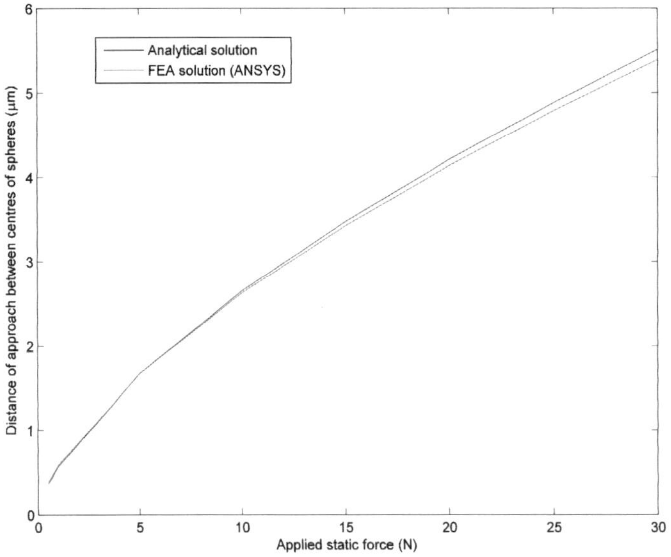

Figure 3. Distance of approach between two spheres in Hertzian contact.

The percentage difference between the analytical solution and the FEA is displayed in figure 4.

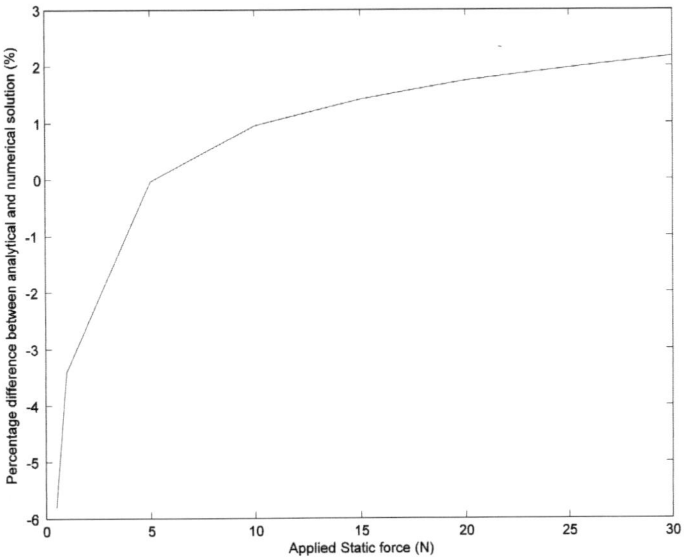

Figure 4. Distance of approach between two spheres in Hertzian contact: percentage difference between analytical solution and FEA.

It can be seen that for this configuration, over the range of applied forces investigated, the agreement between the FEA is within -6% and +2% of the analytical solution. For lower values of the applied force it is likely that issues of numerical tolerance arise, due to the small displacement magnitudes. For larger values of the applied force, an increasingly finer mesh is required around the contact region to minimise the difference between the analytical solution and the FEA. Nevertheless, this level of agreement was considered satisfactory and the case of the dynamic excitation was subsequently investigated.

As described in Section 2, the dynamic excitation case featured a granular chain of six 1 mm diameter chrome steel spheres. The top-most sphere (sphere 1) was excited by a cylindrical piston of 0.2 mm height and of 0.5 mm radius, moving as a rigid body and excited along the *y*-direction by five cycles of a sinusoidal displacement of 0.3 μm magnitude at a frequency of 73 kHz. The final sphere of the chain (sphere 6) was assumed to be in frictionless contact with a rigid support (see figure 5).

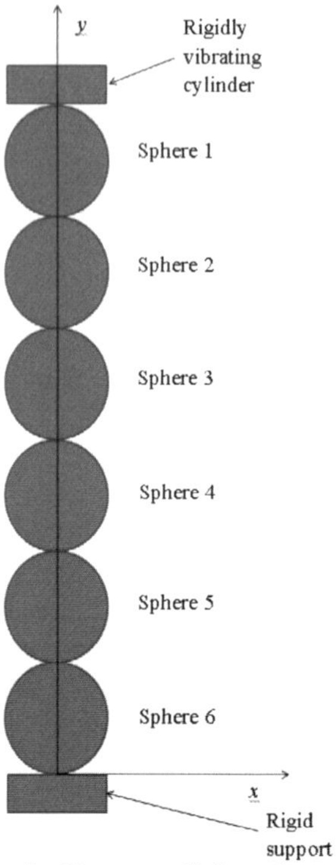

Figure 5. Schematic diagram of six-sphere configuration.

Using a time step of 0.1 μs, a transient analysis was carried out in ANSYS Mechanical, over a time period of 68.5 μs, i.e. for five cycles of the 73 kHz displacement excitation. The Normal Lagrange Formulation was adopted, as described in Section 2. A corresponding discrete mechanics calculation was carried out using the model described in [10], as implemented in [8] and in absence of dissipative effects. The displacement along the *y*-direction at the centre of each sphere is plotted in figures 6a to 6f, corresponding to spheres 1 to 6, respectively.

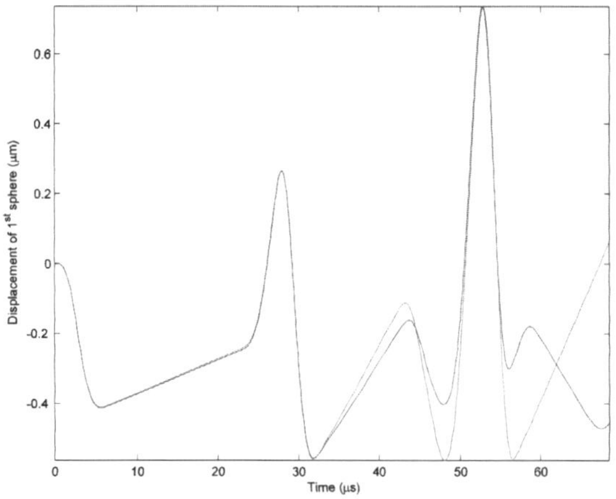

6a Sphere1

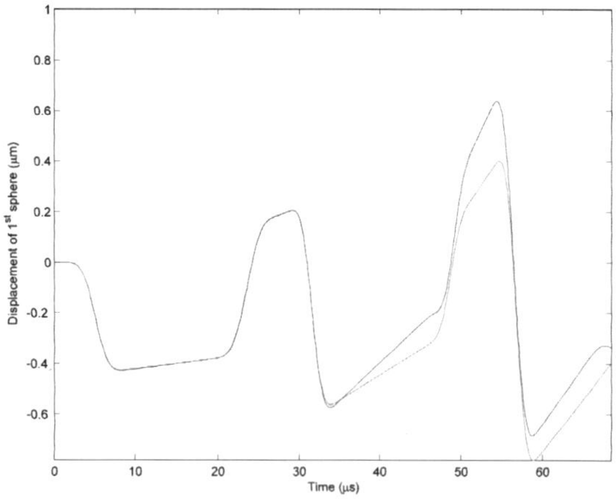

6b Sphere 2

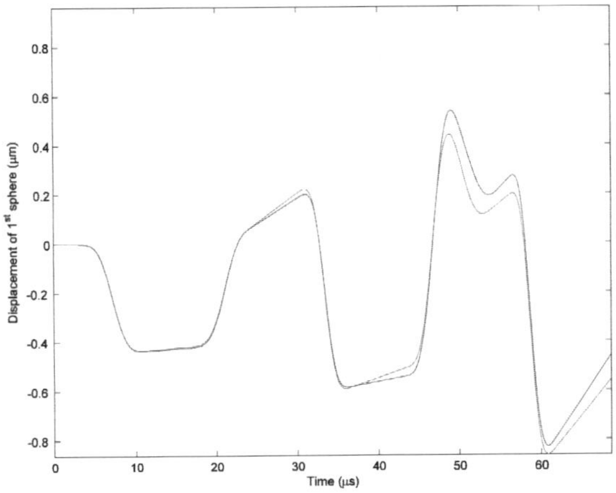

6c Sphere 3

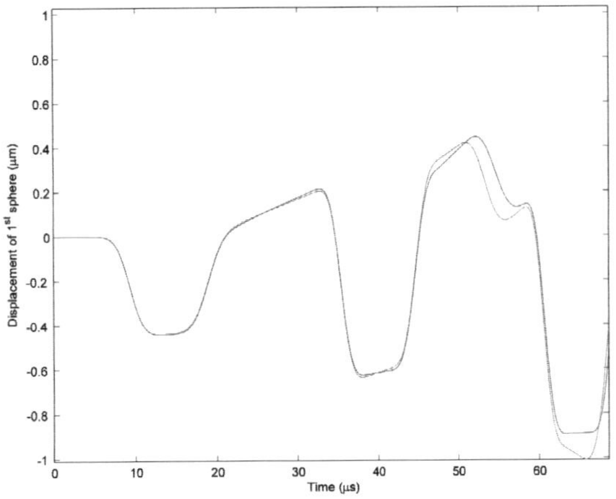

6d Sphere 4

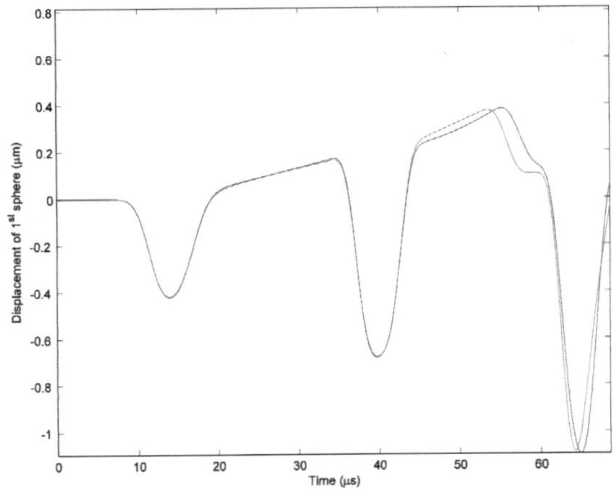

6e Sphere 5

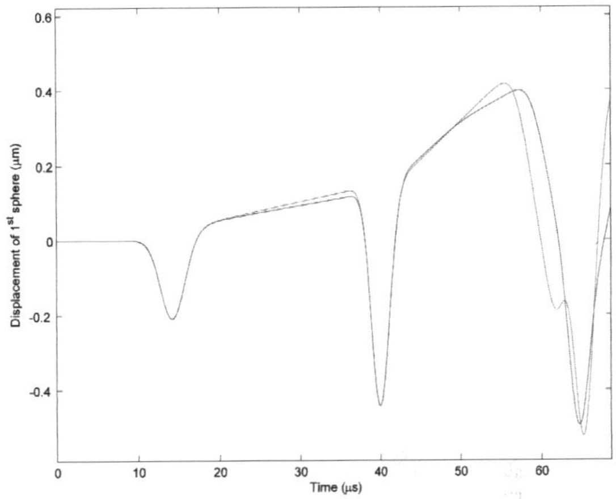

6f Sphere 6

Figure 6 Displacement of the centres of spheres 1 to 6 along the axis of the granular chain as a function of time. The discrete mechanics solution is shown in blue and FEA results are shown in red.

An animation of the total deformation of the beads throughout the five cycles of the 73 kHz displacement excitation can be visualised at the following link: Six-bead granular chain dynamic excitation, where for visualisation purposes, the deformations have been scaled up by a factor of 90.

The above animation clearly shows the wave generated by the impact of the piston on the first bead of the chain. During the excitation cycle, the beads can be seen to lose contact and collide with one another and with the support and the vibrating piston. This effect did not arise in the investigation carried out in [12], due to the nature of the excitation. The impression of penetration of adjacent surfaces in the animation is a result of the large scaling factor. It was verified by post-processing the results that the Normal Lagrange Formulation does indeed preclude penetration at the contact regions.

The graphs in figure 6 show that the output from the discrete mechanics model follows similar trends to the FEA results, in terms of the variation of displacement amplitudes as a function of time. Multiple collisions between the beads are successfully resolved by the FEA. Differences between both models are mainly characterised by time delays in the FEA data, relative to the discrete mechanics waveforms. Indeed, the FEA waveforms appear to lag behind the discrete mechanics results, and it can be seen that the time delay increases as a function of time. Refinement of the numerical solver time step and of the mesh density did not improve the agreement between both models. It is therefore likely that the elastic deformation of the beads is responsible for this. Clearly, if the beads lose contact after

Journal of Physics: Conference Series **684** (2016) 012005 doi:10.1088/1742-6596/684/1/012005

being compressed and do not get the chance to return to their equilibrium spherical shape upon the next impact, this will generate a delay relative to the time of impact estimated using the discrete mechanics model. As successive deformations and impacts occur as the wave propagates up and down the chain, this delay is likely to increase further, and agreement between both models will gradually worsen.

In order to confirm this, the elastic deformation of the beads was investigated. This was done by calculating the cross-sectional distance of each bead along the axis of the granular chain, as a function of time. The result for the first bead is displayed in figure 7.

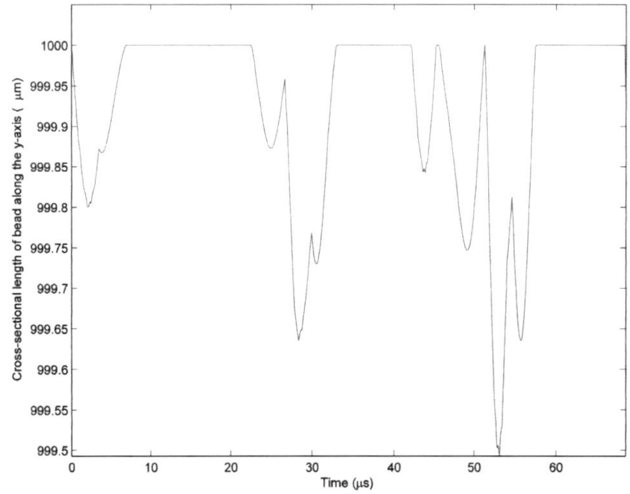

Figure 7 Variation with time of the cross-sectional length of the first bead of the granular chain.

The results in figure 7 clearly show that the bead indeed does not necessarily return to its equilibrium spherical shape upon successive impacts. Additionally, the elastic deformation of the bead can be seen to be of the order of the piston maximum displacement amplitude, which is 0.3 μm. This observation, which is also valid for the other beads of the chain, highlights a limitation of discrete mechanics models when investigating the dynamics of granular chains.

4. Conclusions

This work constitutes a preliminary investigation towards designing novel devices for biomedical ultrasound applications, whose transduction process will rely on non-linear phenomena arising from the dynamic excitation of granular chains.

A finite element model was developed in ANSYS Mechanical to simulate the dynamic behaviour of a granular chain consisting of six spherical beads of 1 mm diameter. The first bead of the chain was excited by a rigidly vibrating cylindrical piston, undergoing a 0.3 μm amplitude sinusoidal displacement, along the direction of the axis of the chain at a frequency of 73 kHz and for five cycles. Initially, it was ensured that Hertz' contact law was satisfied for the static case of two pre-compressed spheres. Subsequently, axial displacements at the bead centres obtained from the FEA of the dynamic excitation of the chain were compared with results from the well-known discrete mechanics model. Good overall agreement between both models was observed. There were, however, discernible differences which were attributed in part to the elastic deformation of the beads, which is not accounted for in the discrete mechanics model when the beads separate.

In [12], it was reported that plastic deformations of the beads could constitute a source of dissipation and lead to differences between experimental data and results from discrete mechanics models. The study reported in this paper investigates the dynamics of a granular chain with excitation and boundary conditions which cause the beads to separate and collide with one another, which was not the case in [12]. This present study therefore highlights a potential additional source of disagreement between experimental results and discrete mechanics models, in the dynamic analysis of granular chains.

Acknowledgements

The authors gratefully acknowledge funding from the Engineering and Physical Sciences Research Council (UK) via grant number EP/K030159/1. Useful discussions with Xiaoping Jia and with Ryan Musson are also acknowledged.

References

[1] Cai L, Yang J, Rizzo P, Ni X and Daraio C 2013 Propagation of highly nonlinear solitary waves in a curved granular chain Granul. Matter 15 357–366.
[2] Nesterenko V F 1983 Propagation of nonlinear compression pulses in granular media J. Appl. Mech. Tech. Phys. 24 733–43.
[3] Sen S, Hong J, Avalos E, and Doney R 2008 Solitary waves in a granular chain, Physics Reports 462 21−66.
[4] Nettel S, Wave Physics, Oscillations-Solitons-Chaos, Fourth Edition, Springer, 2009.
[5] Lazaridi A N and Nesterenko V F 1985 Observation of a new type of solitary waves in a one dimensional granular medium J. Appl. Mech. Tech. Phys. 26(3) 405–408.
[6] Coste C, Falcon E, Fauve S 1997 Solitary waves in a chain of beads under Hertz contact, Physical Review E 56 6104−6117.
[7] Spadoni A and Daraio C 2010 Generation and control of sound bullets with a nonlinear acoustic lens Proc. Natl. Acad. Sci. 107 7230−7234.
[8] Hutchins D A, Yang J, Akanji O, Thomas P J, Davis L A J, Freear S, Harput S, Saffari N and Gélat P 2015 Evolution of ultrasonic impulses in chains of spheres using resonant excitation *Eur. Phys. Lett.* 109(5) 54002.
[9] Gavrilov L R, Hand J W 2000 A theoretical assessment of the relative performance of spherical phased arrays for ultrasound surgery IEEE T. Ultrason. Ferr. 47 125–139.
[10] Lydon J, Jayaprakash K R, Ngo D, Starosvetsky Y, Vakakis A F and Daraio C 2013 Frequency bands of strongly nonlinear homogeneous granular systems, Physical Review E, 88 012206(9)
[11] Yastrebov V A 2014 Numerical Methods in Contact Mechanics Wiley-ISTE.
[12] Musson R W and Carlson W 2014 Simulation of solitary waves in a monodisperse granular chain using COMSOL multiphysics: localized plastic deformation as a dissipation mechanism Granul. Matter 16 543–550.
[13] COMSOL Multiphysics Structural Mechanics Module User's Guide, Version 3.5a, COMSOL, Inc., (2008).
[14] Hertz H R 1881 On the contact of elastic solids J. Reine Angew. Math. 92 156–171.
[15] ANSYS Mechanical User's Guide, Version 15.0, ANSYS Inc., (2013).

Use of shear waves for diagnosis and ablation monitoring of prostate cancer: a feasibility study

A Gomez[1], G Rus[2] and N Saffari[1]

[1] UCL Mechanical Engineering, University College London, Torrington Place, London WC1E 7JE, UK

[2] Department of Structural Mechanics, University of Granada, Politécnico de Fuentenueva, Granada 18071, Spain

E-mail: aj.gomez@ucl.ac.uk

Abstract. Prostate cancer remains as a major healthcare issue. Limitations in current diagnosis and treatment monitoring techniques imply that there is still a need for improvements. The efficacy of prostate cancer diagnosis is still low, generating under and over diagnoses. High intensity focused ultrasound ablation is an emerging treatment modality, which enables the non-invasive ablation of pathogenic tissue. Clinical trials are being carried out to evaluate its long-term efficacy as a focal treatment for prostate cancer. Successful treatment of prostate cancer using non-invasive modalities is critically dependent on accurate diagnostic means and is greatly benefited by a real-time monitoring system. While magnetic resonance imaging remains the gold standard for prostate imaging, its wider implementation for prostate cancer diagnosis remains prohibitively expensive. Conventional ultrasound is currently limited to guiding biopsy. Elastography techniques are emerging as a promising real-time imaging method, as cancer nodules are usually stiffer than adjacent healthy prostatic tissue. In this paper, a new transurethral approach is proposed, using shear waves for diagnosis and ablation monitoring of prostate cancer. A finite-difference time domain model is developed for studying the feasibility of the method, and an inverse problem technique based on genetic algorithms is proposed for reconstructing the location, size and stiffness parameters of the tumour. Preliminary results indicate that the use of shear waves for diagnosis and monitoring ablation of prostate cancer is feasible.

1. Background

Prostate Cancer (PCa) is the most common cancer in men in the UK. It is also the second cause of cancer death after lung cancer. It represents around 13% of all cases of cancer and accounts for 7% of all UK cancer deaths [1]. The incidence of PCa increases with age, affecting mainly men over 50 years of age [2]. The increase in longevity and awareness of the disease will lead to more men requesting screening, which will in turn increase the number of patients diagnosed with PCa in the future [3].

Prostate carcinoma is often suspected when the serum Prostate-Specific Antigen (PSA) is elevated or an abnormal digital rectal examination is noted. However, PSA screening leads to a substantial number of unnecessary biopsies in patients with no or indolent cancer, who do not need immediate treatment [4]. Histopathological evaluation of systematic biopsy cores is used to confirm or rule out cancer, and is normally carried out with conventional transrectal ultrasound as a guidance tool. Despite the low specificity of PSA testing and the uncertainty after negative systematic biopsies, these techniques remain the standard for PCa diagnosis.

Current research is investigating whether modern imaging techniques can identify the site of the tumour. There are two main streams that are showing promising preliminary results: Magnetic Resonance Imaging (MRI) and Elastography techniques.

Multi-parametric MRI (mpMRI) combines T2-weighted imaging with functional sequences. It has become an important modality for tumour detection and staging [5,6], however, mpMRI performance varies depending on which combination of features is selected [7]. Additional limitations include cost, limited availability, contra-indication to MRI and contrast agents, and the fact that the very large majority of biopsies are ultrasound guided [8].

Prostate Elastography is an emerging imaging modality, which consists of the evaluation of prostate stiffness. Analogous to the cellular processes of wound repair, it is generally believed that normal tissue stroma responds in an effort to repair damage due to carcinoma cell invasion [9]. It has been concluded that the stromal reaction is also characterized by elevated collagen deposition [10]. Since increasing collagen deposition leads to an increase in the tumour rigidity, this suggests that quantitative stiffness estimations may prove to be an effective biomarker for assessing PCa grade and identification of more aggressive cancers [11].

Real-Time Elastography is available on some ultrasound systems for prostate imaging [12,13] and a few other techniques are currently being developed [13]. There are two main approaches commercially available:

- Quasi-Static (or Strain) Elastography (SE);
- Shear Wave Elastography (SWE).

SE of the prostate is based on the comparative analysis of tissue deformation before and after applying a slight static mechanical compression through the rectal wall. Stiffer tissue experiences less deformation than normal tissue. No quantitative elasticity analysis is available. Relative changes in the strain rate between zones work as a guide to suspect pathologic nodules existence [14]. This technique is commercially available for many clinical ultrasonic platforms [12]. SE limitations include:

- The lack of uniform compression over the entire gland;
- The intra and inter-operator dependency;
- Penetration issues in large prostate glands;
- Artefacts due to slippage of the compression plane.

SWE in PCa detection has been tested mainly by the Aixplorer ultrasound system transrectally (SuperSonic Imagine, Aix-en-Provence, France). In Transrectal-SWE (TR-SWE) by the Aixplorer system, acoustic radiation force produces a shear wave in the shape of a cone with a small inclination, which travels away from the pushing beam. An ultrafast scanner allows shear waves to be followed in real time in 2D, and echo tracking produces a displacement recording from which a small map of elasticity can be created [15]. Spatial resolution is worse than in SE imaging but the elastograms are quantitative.

Recent studies on PCa diagnosis using TR-SWE have shown very promising results [16–20]. The use of a 35 kPa threshold for Young's modulus to separate lesions from normal tissue in the Peripheral Zone may provide additional information for PCa detection and biopsy guidance, enabling a substantial reduction in the number of biopsies [16,17]. TR-SWE limitations include:

- The pressure artefacts induced by the transducer, as the end-fire design of the probe requires bending to image the mid prostate and apex;
- The slow frame rate, i.e. one image per second;
- The limited size of the region of interest (concretely only half of the prostate is covered);
- The delay to stabilize the signals at each acquisition plane;
- The signal attenuation in large prostates making the evaluation of the anterior Transitional Zone difficult or impossible [8].

The aim of this study is to evaluate the feasibility of using a Transurethral-SWE (TU-SWE) approach for diagnosis and High Intensity Focused Ultrasound (HIFU) ablation monitoring of PCa. This approach encompasses both the transmission and the detection of the shear wave propagation. Shear waves are propagated into the prostate as the result of applying torsional stress on the urethra wall at a localized point. If stiff lesions are presented in the prostate, the wavefront would suffer reflections due to the change in acoustic impedance. This reflection would travel back to the urethra where it could be detected (Figure 1). The parameters which define the tumour (location, size and stiffness) can be reconstructed from those recovered signals.

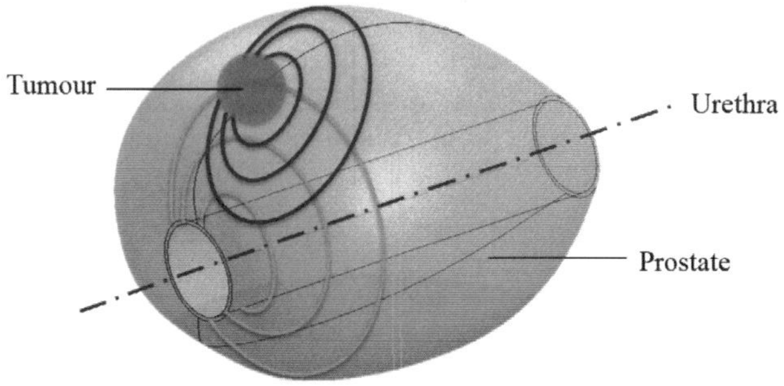

Figure 1. Conceptual idealization of the TU-SWE approach for diagnosis and ablation monitoring of PCa

The capability of transmitting shear waves into the prostate by mechanical contact through the urethra wall was evaluated in phantoms and *in vivo* canine experiments, by using a mechanical transurethral actuator as excitation force and MRI Elastography for measuring the displacement [21,22].

HIFU ablation is a promising focused technique for treating PCa, currently under clinical evaluation in several clinics around the world. As with other focused therapies it requires that the treated region be controlled and monitored to avoid damage of the surrounding healthy tissue. The TU-SWE approach allows the monitoring of HIFU ablation from the rectum, since the ultrasonic beam itself generates an alteration of the stiffness at the focal point and surroundings as the temperature increases during the surgery [23,24].

Inherent advantages of the TU-SWE approach proposed are listed below:

- The Transitional Zone and the Central Zone of the prostate (Figure 2) remain less accessible when using techniques that operate from the rectal passage. The urethral approach makes these zones more reachable. The Central Zone presents the lowest rate of PCa, but these cancers tend to be more aggressive and more likely to invade the seminal vesicles [25]. The TR-SWE studies are limited to the Peripheral Zone since this approach makes it difficult to reach the anterior zones of the prostate [16]. On the other hand, the reduction of tissue path is directly connected to the diminution of attenuation, which is a crucial matter in shear wave propagation.
- The possibility of using higher frequencies (above 500 Hz) than other techniques. This fact is related to the reduction in attenuation, which allows a slight increase of the excitation frequency, and therefore the spatial resolution, thus improving the imaging quality and the capacity of detecting smaller cancer nodules.
- Current techniques present difficulties in scanning the whole gland in a short time. TR-SWE covers only half of the prostate for every 2D scan. Although SE can cover complete 2D sections of the prostate, it presents problems covering the whole gland due to the challenge of keeping the pressure constant at every scan. Due to the geometrical configuration of the proposed TU-SWE approach, the whole gland volume may be covered simultaneously, permitting the opportunity of 3D real-time monitoring technique.

- In comparison with techniques that use acoustic radiation force, lower levels of energy are expected, and therefore lower ultrasound thermal and mechanical indices.
- The use of the urethra as a channel for diagnosis keeps the rectal passage free for transrectal therapies such as HIFU ablation.

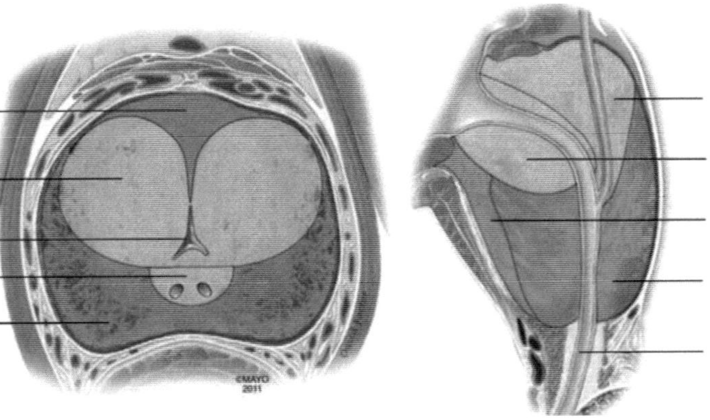

Figure 2. McNeal's anatomic division of the prostate. Source: Mayo Foundation for Medical Education and Research [26].

In this paper, a prostate-like medium is modelled using the Finite-Difference Time Domain (FDTD) method to investigate the PCa detection sensitivity of the proposed approach. An Inverse Problem (IP) based on genetic algorithms was developed for reconstructing the parameters that locate and size prostate nodules. Finally, the results from the reconstruction were analyzed to demonstrate the feasibility of the TU-SWE approach for diagnosis of PCa.

2. Methodology

2.1. Finite-Difference Time Domain model

A FDTD method is developed for simulating the propagation of shear waves into the prostate from the urethra. The FDTD method has been extensively used for modelling different physics phenomena in solid materials, mainly in electromagnetism and geophysics, but also for wave propagation in elastic and viscoelastic media [27,28]. Two aims are pursued with this model. One is to study, in a simple way, the feasibility of reflection detection in the PCa application. The other is to feed the reconstruction method, obtaining potential solutions which are tested iteratively in the inverse procedure.

The mechanical response of soft tissues is well known for a viscoelastic behaviour. In this study, an elastic model has been considered sufficient due to the absence of consensus about mechanical models and values for viscous parameters of prostatic tissue.

A cylindrical coordinate system is selected for establishing the discretization grid. The simplified prostate-like model is defined geometrically as a hollow cylinder, where the inner passage corresponds to the urethra (Figure 3).

Although the equations of motion can be formulated in several ways, a velocity-stress formulation was chosen [29], where time distributions of velocity and stress at spatial points are propagated along a regular grid to which material properties and the excitation source are assigned. Generally, a system of nine first-order hyperbolic equations are needed. Shear waves are generated by the application of torsional forces in the urethra wall in an axisymmetric configuration, and thus, together with the simplification of using a two dimensional approach, reduces the complexity to only one velocity component v_θ. The plane selected (in red colour Figure 3) contains the axis. The remaining three first-order hyperbolic equations (1-3) are expressed in terms of the stress tensor σ and velocity v_θ for an elastic medium. As a result of the simplification, the circular tumour in the plane becomes a torus in 3D.

Although this geometrical distribution is not representative of PCa, it can be considered acceptable for a proof of principle approach.

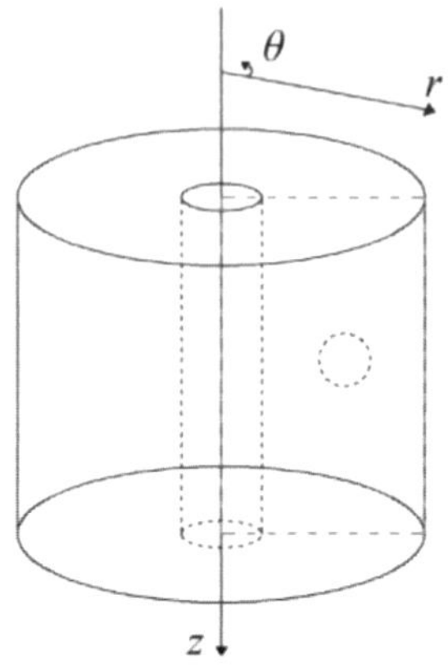

Figure 3. Simplified geometry for the model. 2D section of study in red. Spherical tumour.

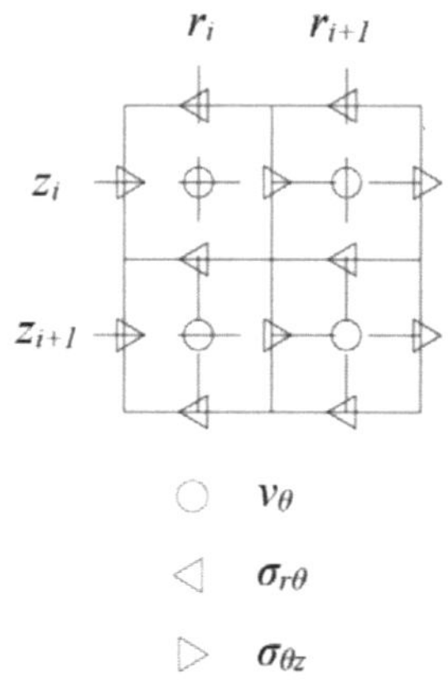

Figure 4. Diagram of the staggered grid discretization with positions of the variables.

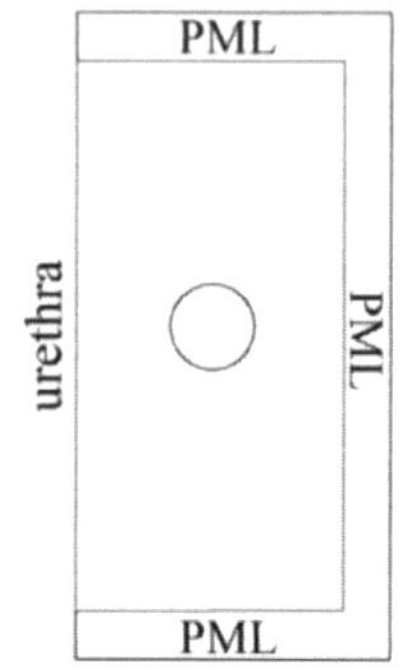

Figure 5. Spatial distribution of the model. Main domain in red. PML surrounding it. Urethra at left side.

$$\rho \dot{v}_\theta = \frac{\mathrm{d}\sigma_{r\theta}}{\mathrm{d}r} + \frac{\mathrm{d}\sigma_{\theta z}}{\mathrm{d}z} + \frac{2}{r}\sigma_{r\theta} \tag{1}$$

$$\dot{\sigma}_{r\theta} = \mu\left(\frac{\mathrm{d}v_\theta}{\mathrm{d}r} - \frac{v_\theta}{r}\right) \tag{2}$$

$$\dot{\sigma}_{\theta z} = \mu\frac{\mathrm{d}v_\theta}{\mathrm{d}z} \tag{3}$$

The mass density ρ and the shear modulus μ vary depending on whether they belong to the background tissue or to the cancer nodule. Spatial discretization is achieved with the staggered grid illustrated in Figure 4, i.e., the velocity component is specified at grid positions that are offset by a half-step from the corresponding stress components. Time is uniformly sampled via $t=n\Delta t$ for an integer n and interval Δt. Similarly, space is uniformly sampled, with $r=i\Delta r$ and $z=j\Delta z$ for integers i, j.

To propagate spatial quantities in time, the classical time-staggering approach is often used to estimate stress and velocity in alternating time intervals. In this case, an alternative method has been chosen, computing both magnitudes at each time interval starting from an initial situation at rest [27].

All temporal and spatial derivatives in equations (1), (2) and (3) are discretized using the following differencing schemes:

$$\dot{g}\Big|_{r_i,t_{n+1}} = \frac{\mathrm{d}g}{\mathrm{d}t}\Big|_{r_i,t_{n+1}} = \frac{g^{n+1}(i) - g^n(i)}{\Delta t} + \mathrm{O}(\Delta t) \tag{4}$$

$$g_{,r}\Big|_{r_i,t_n} = \frac{\mathrm{d}g}{\mathrm{d}x}\Big|_{r_i,t_n} = \frac{g^n\left(i+\frac{1}{2}\right) - g^n\left(i-\frac{1}{2}\right)}{\Delta r} + \mathrm{O}(\Delta r^2) \tag{5}$$

To reduce numerical dispersion we choose a minimum rate of 20 spatial intervals per wavelength λ as described in equation (6) although lower rates can be found in the literature. The temporal interval satisfies the stability condition (7) obtained from the Von Neumann condition [30].

$$20\max(\Delta r,\Delta z) < \lambda_{\min} \tag{6}$$

$$\max(c_s)\Delta t < \left(\frac{1}{\Delta r^2}+\frac{1}{\Delta z^2}\right)^{-1/2} \tag{7}$$

The rotational excitation is introduced in the model by setting v_θ at the corresponding points in the grid within the urethra wall where the torsion is applied.

In order to avoid reflections at outer boundaries of the domain, a Perfectly Matched Layer (PML) absorber is incorporated surrounding the main domain, excepting the urethra wall (Figure 5), using an adaptation for cylindrical coordinates [31].

2.2. *Inverse Problem based on Genetic Algorithms*

A way to solve an IP is to use the forward problem to find a solution for a set of parameters that minimizes the difference between the predicted and the actual measurements by using an optimization algorithm [32]. For this study a genetic algorithm is proposed for the optimization. The tumour parameters are contained in a vector named p_c. These parameters are: the radial r_c and depth z_c coordinates of the centroid of the nodule, the diameter ϕ_c and the stiffness μ_c of the tumour.

Considering that this work is at a preliminary stage, experimental results are not yet available. To study the suitability of the proposed reconstruction method, pseudo-experimental signals u_{exp} were generated by adding white noise to signals simulated using the FDTD model.

The optimization algorithm will minimize the discrepancy between u_{exp} and the numerically predicted trial response $u(p_c)$, which is a function of the four-dimensional hyperspace formed by the four tumour parameters. To quantify the discrepancy, a cost functional f is designed, defined as the squared Euclidean distance between both measurements (8) [33].

$$f(p_c) = \left\| u_{\exp} - u(p_c) \right\|_2^2 \tag{8}$$

3. Results

3.1. *Numerical model results*

A numerical simulation was carried out by using the FDTD method described in Section 2.1. Figure 6 depicts the selected set-up. Note that the problem has been simplified after taking advantage of the axisymmetry of the cylindrical system shown in Figure 3.

The values corresponding to the labeled variables in Figure 6 are listed in Table 1. Those values and proportions are in agreement with those that can be found in a real human prostate. A rounded shape is selected to represent the cancerous nodule using an arbitrary, but realistic, value of 4 mm of diameter. The excitation source is located on the urethra wall, at centered depth. A set of 18 receivers is uniformly distributed for recording the signals coming from the interior of the domain.

Table 2 shows the values for mechanical parameters of both healthy and cancerous prostate, as well as the settings of the excitation source, which is implemented as a Gaussian modulated wave with central frequency of 1 kHz. Shear modulus values have been chosen in agreement with those found in the literature [16,17].

Table 3 lists the parameters for the numerical discretization. In order to guarantee an adequate visual resolution of the simulation, the discrete spatial values go beyond the stability condition described by equation (6), resulting in a total of 731,600 elements. The Von Neumann condition defined by equation (7) is applied to calculate the temporal discretization interval.

The simulation was performed using MATLAB® in conjunction with the Parallel Computing Toolbox™ (Release 2014b, MathWorks, Natick, United States). Run times lasted 406 seconds on a quad-core 3.60 GHz, 16 GB RAM, desktop computer.

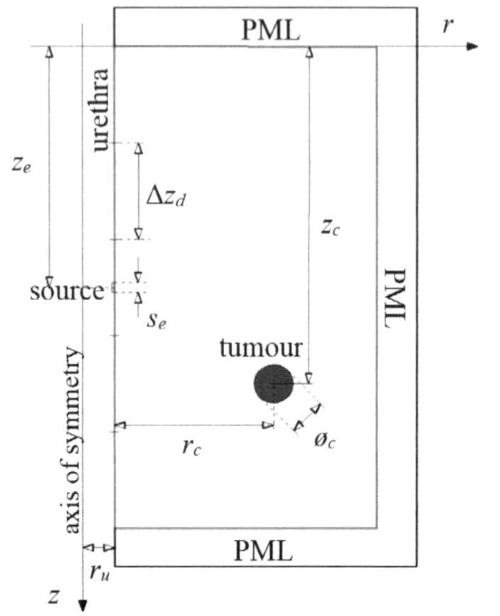

Figure 6. Schematic of set-up for numerical simulation. Tumour in red. Dimensions are described in Table 1.

Table 1. Dimension values for set-up of the numerical simulation.

Variable	Description	Value (mm)
r_d	Radial dimension of the domain	23.50
z_d	Depth dimension of the domain	50.00
r_u	Radius of urethra	3.25
z_e	Depth coordinate of excitation source	25.00
s_e	Thickness of excitation source	1.00
r_c	Radial coordinate of the tumour	14.00
z_c	Depth coordinate of the tumour	35.00
$\emptyset_c$	Diameter of the tumour	4.00
Δz_r	Distance between receivers	2.38

Table 2. Values for mechanical parameters and excitation [16,17]

Variable	Description	Value
ρ_h	Density of healthy tissue	1000 kg m^{-3}
μ_h	Shear modulus of healthy tissue	6.00 kPa
ρ_c	Density of cancerous tissue	1000 kg m^{-3}
μ_c	Shear modulus of cancerous tissue	18.00 kPa
f_c	Central frequency	1.00 kHz
t_T	Total time of simulation	25 ms
a	Source amplitude	10.00 μm

Table 3. Discretization parameters for FDTD simulation

Variable	Description	Value
Δr	Radial dimension interval	50 μm
Δz	Depth dimension interval	50 μm
Δt	Time interval	8.33 μs
n_{PML}	Number of PML elements	120

Figure 7 shows four snapshots of the simulated transient propagation of shear waves in a prostate-like medium containing a cancerous lesion, at different times when the wavefront hits and propagates across the tumour. It can be clearly seen how the reflections are generated traveling back to the urethra wall, as well as the ongoing wavefront accelerates while propagating across the tumour due to the higher velocity of the shear waves, 4.24 ms^{-1} against 2.45 ms^{-1} in the healthy tissue. A second reflection coming from the bottom of the tumour can be also visualized. The recordings at the receivers during the simulation are displayed in Figure 8.

3.2. Reconstruction IP-AG method results

In order to obtain a proof of feasibility for the IP-GA method, an example reconstruction is shown in this section.

As mentioned in Section 2.2, a simulated signal generated by the FDTD model with added white noise was used as an experimental signal to be reconstructed by the IP-GA method. The amplitude of the white noise signal corresponded to 10% of the root-mean-square of the simulated signal amplitude. The set-up for the pseudo-experimental signal was the same as the one in Section 2.1 (see Figure 6 and Tables 1 and 2). Table 4 shows the values of the parametrization. In this case, a coarser grid was chosen in order to speed up the iteration process and reduce the computational time. The pseudo-experimental signal was detected by a receiver located 12.5 mm below the source on the urethra wall. The reconstruction algorithm was implemented using MATLAB® and the Parallel Computing Toolbox™ (Release 2014b, MathWorks, Natick, United States), lasting for 1204 seconds on the computer mentioned above. A population of 10 candidate solutions was selected, to be updated during 13 generations.

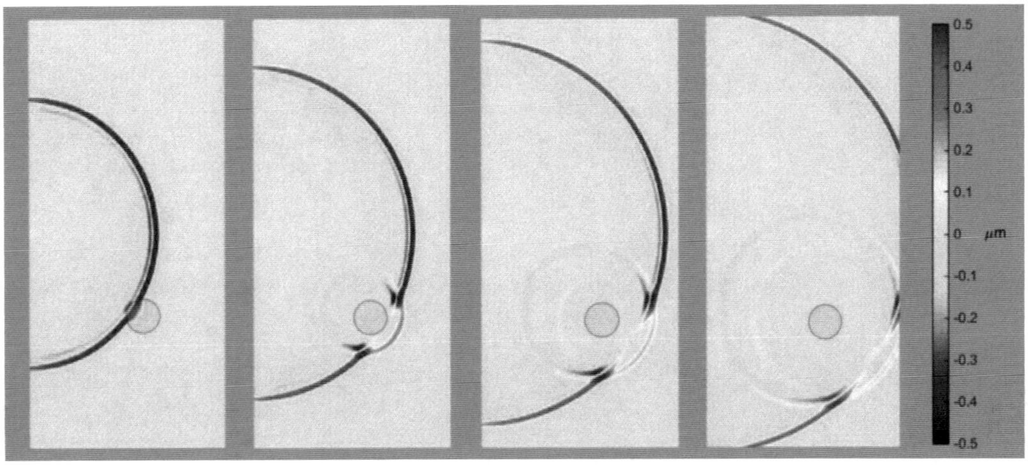

Figure 7. Snapshots of the FDTD simulation at 7.2, 8.7, 9.8 and 11.3 ms. The colour bar represents the amplitude of the displacement. The 4 mm rounded tumour is highlighted.

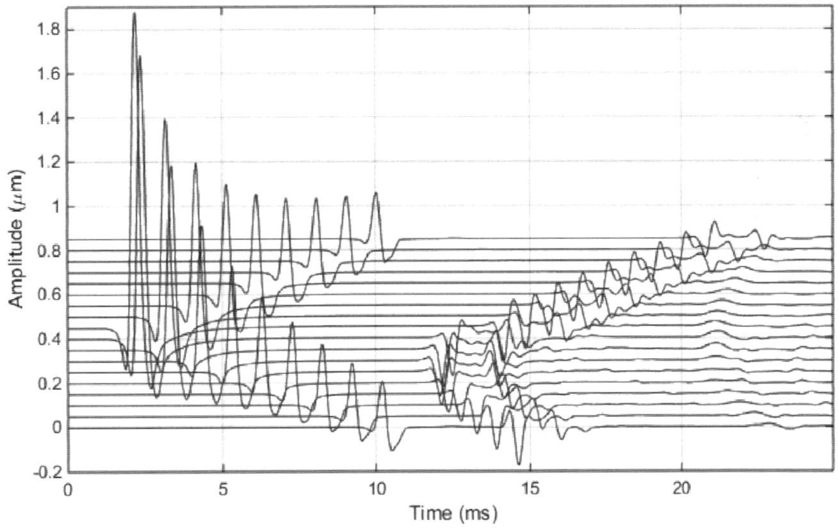

Figure 8. Recovered displacement signals at the 18 receivers. Distance between receivers 2.38 mm, with a 7.14 mm gap between receiver 9 and 10, where the excitation source is positioned.

Figure 9 depicts the reconstructed signal over the pseudo-experimental one. Recovered values for the tumour parameters and reconstruction errors are shown in Table 5. The errors were calculated over the search range established for each parameter. The algorithm showed good agreement between the reconstruction and the pseudo-experimental measurement.

Table 4. Discretization parameters for IP-AG testing

Variable	Description	Value
Δr	Radial dimension interval	150 μm
Δz	Depth dimension interval	150 μm
Δt	Time interval	17.32 μs
n_{PML}	Number of PML elements	40

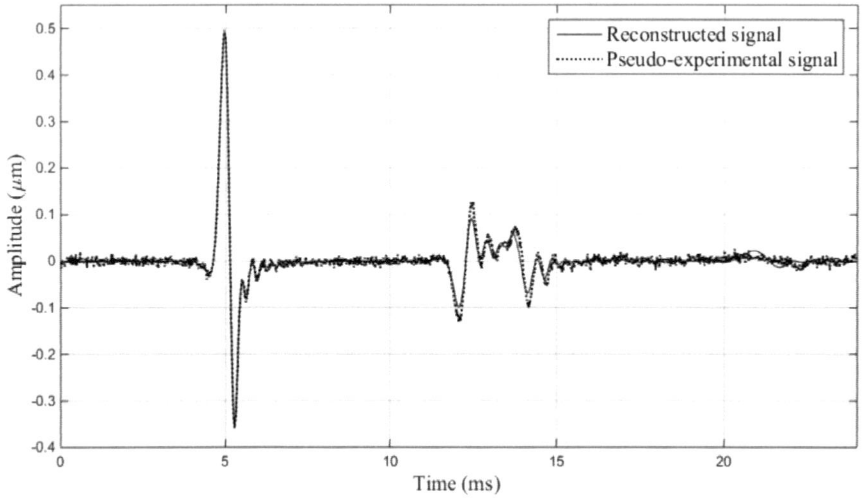

Figure 9. Reconstructed signal by using the IP-AG method (blue solid line) over the pseudo-experimental signal (black dotted line).

Table 5. Reconstructed values for the tumour parameters and reconstruction errors.

Parameter	Description	Original value	Reconstructed Value	Search Range	Error
r_c	Radial coordinate of the tumour	14.00 mm	13.7 mm	4 - 24 mm	3.10 %
z_c	Depth coordinate of the tumour	35.00 mm	36.98 mm	26 - 44 mm	3.96 %
ϕ_c	Diameter of the tumour	4.00 mm	3.63 mm	2 - 8 mm	6.16 %
μ_c	Shear modulus of cancerous tissue	18.00 kPa	16.61 kPa	16 - 21 kPa	9.56 %

4. Conclusion

A novel TU-SWE approach for diagnosis and ablation monitoring of PCa has been presented in this paper. It is based on the emission, detection and computational treatment of shear waves propagated into the prostate.

In order to study the feasibility of this approach, a FDTD model was developed to analyze the behavior of shear waves in a medium representative of prostatic tissue. Although the medium was assumed to be homogeneous and elastic, this was considered sufficient to simulate the mechanical behavior of the human prostate to a first approximation. The presence of cancerous nodules, which are usually stiffer than the surrounding healthy tissue, was also modelled. Reflections were generated when the wavefront hit the tumour, traveling back to the urethra where they were detected, demonstrating proof of principle and warranting further studies.

The IP-AG reconstruction method was tested by using the FDTD model for generating a pseudo-experimental signal and also as a forward model for searching the solutions within the IP algorithm. The results from the test showed good agreement between the pseudo-experimental signal and the reconstructed one, with an admissible level of error in the recovery of tumour parameters.

In further studies, a more realistic 3D model will be required, taking into account the heterogeneity and mechanical complexity of the human prostate. Experimental tests in prostate phantoms will be also required. In the IP ambit, it is of interest to study the probability of producing a false positive diagnosis, as well as how the response of the system changes with increasing levels of noise.

5. Acknowledgements

The authors would like thank P. Gelat for constructive and illuminating discussions. This research was supported by the Ministry of Education DPI2014-51870-R and DPI2010-17065, and Junta de Andalucía for projects P11-CTS-8089 and GGI3000IDIB. First author was supported by Talentia Scholarship C2012F1-75146405T-1, Agencia del Conocimiento Junta de Andalucía.

References

[1] Cancer Research UK 2014 Prostate cancer Key Stats *Cancer Res. UK*

[2] The Prostate Cancer Charity UK 2015 What is my risk of prostate cancer? *Prostate Cancer UK*

[3] Abdellaoui A, Iyengar S and Freeman S 2011 Imaging in prostate cancer *Futur. Oncol.* 679–91

[4] Kelloff G J, Choyke P and Coffey D S 2009 Challenges in clinical prostate cancer: role of imaging *AJR. Am. J. Roentgenol.* **192** 1455–70

[5] Lemaitre L, Puech P, Poncelet E, Bouyé S, Leroy X, Biserte J and Villers A 2009 Dynamic contrast-enhanced MRI of anterior prostate cancer: Morphometric assessment and correlation with radical prostatectomy findings *Eur. Radiol.* **19** 470–80

[6] Lim H K, Kim J K, Kim K a. and Cho K S 2009 Prostate cancer: Apparent diffusion coefficient map with T2-weighted images for detection - A multireader study *Radiology* **35** 100–100

[7] Langer D L, Van Der Kwast T H, Evans A J, Trachtenberg J, Wilson B C and Haider M a. 2009 Prostate cancer detection with multi-parametric MRI: Logistic regression analysis of quantitative T2, diffusion-weighted imaging, and dynamic contrast-enhanced MRI *J. Magn. Reson. Imaging* **30** 327–34

[8] Correas J M, Drakonakis E, Isidori a. M, Hélénon O, Pozza C, Cantisani V, Di Leo N, Maghella F, Rubini a., Drudi F M and D'ambrosio F 2013 Update on ultrasound elastography: Miscellanea. Prostate, testicle, musculo-skeletal *Eur. J. Radiol.* **82** 1904–12

[9] Tuxhorn J a, Ayala G E and Rowley D R 2001 Reactive stroma in prostate cancer progression. *J. Urol.* **166** 2472–83

[10] Tuxhorn J a, Ayala G E, Smith M J, Smith V C, Dang T D and Rowley D R 2002 Reactive Stroma in Human Prostate Cancer : Induction of Myofibroblast Phenotype and Extracellular Matrix Remodeling Reactive Stroma in Human Prostate Cancer : Induction of Myofibroblast Phenotype and Extracellular **8** 2912–23

[11] Burns-Cox N, Avery N C, Gingell J C and Bailey a J 2001 Changes in collagen metabolism in prostate cancer: a host response that may alter progression. *J. Urol.* **166** 1698–701

[12] Bamber J, Cosgrove D, Dietrich C F, Fromageau J, Bojunga J, Calliada F, Cantisani V, Correas J-M, D'Onofrio M, Drakonaki E E, Fink M, Friedrich-Rust M, Gilja O H, Havre R F, Jenssen C, Klauser a S, Ohlinger R, Săftoiu A, Schaefer F, Sporea I and Piscaglia F 2013 EFSUMB guidelines and recommendations on the clinical use of ultrasound elastography. Part 2: Clinical applications. *Ultraschall Med.* **34** 238–53

[13] Doherty J, Trahey G, Nightingale K and Palmeri M 2013 Acoustic radiation force elasticity imaging in diagnostic ultrasound *IEEE Trans. Ultrason. Ferroelectr. Freq. Control* **60** 685–701

[14] Ophir J, Alam S K, Garra B S, Kallel F, Konofagou E E, Krouskop T, Merrit C R B, Righetti R, Souchon R, Srinivasan S and Varghese T 2002 Elastography : Imaging the Elastic Properties

of Soft Tissues with Ultrasound *J. Med. Ultrason.* **29** 155–71

[15] Bercoff J, Tanter M and Fink M 2004 Supersonic Shear Imaging : A New Technique for Soft Tissue Elasticity Mapping **51** 396–409

[16] Barr R G, Memo R and Schaub C R 2012 Shear wave ultrasound elastography of the prostate: initial results *Ultrasound Q.* **28** 13–20

[17] Correas J M, Tissier A-M, Khairoune A, Vassiliu V, Méjean A, Hélénon O, Memo R and Barr R G 2015 Prostate Cancer : Diagnostic Performance of Real-time Shear-Wave Elastography *Radiology* **275** 280–9

[18] Ahmad S, Cao R, Varghese T, Bidaut L and Nabi G 2013 Transrectal quantitative shear wave elastography in the detection and characterisation of prostate cancer *Surg. Endosc.* **27** 3280–7

[19] Woo S, Kim S Y, Lee M S, Cho J Y and Kim S H 2015 Shear wave elastography assessment in the prostate: an intraobserver reproducibility study *Clin. Imaging* **39** 484–7

[20] Boehm K, Salomon G, Beyer B, Schiffmann J, Simonis K, Graefen M and Budaeus L 2015 Shear Wave Elastography for Localization of Prostate Cancer Lesions and Assessment of Elasticity Thresholds: Implications for Targeted Biopsies and Active Surveillance Protocols *J. Urol.* **193** 794–800

[21] Chopra R, Arani A, Huang Y, Musquera M, Wachsmuth J, Bronskill M and Plewes D 2009 In vivo MR elastography of the prostate gland using a transurethral actuator. *Magn. Reson. Med.* **62** 665–71

[22] Arani A, Plewes D and Chopra R 2011 Transurethral prostate magnetic resonance elastography: prospective imaging requirements. *Magn. Reson. Med.* **65** 340–9

[23] Arnal B, Pernot M and Tanter M 2011 Monitoring of thermal therapy based on shear modulus changes: I. Shear wave thermometry *IEEE Trans. Ultrason. Ferroelectr. Freq. Control* **58** 369–78

[24] Arnal B, Pernot M and Tanter M 2011 Monitoring of thermal therapy based on shear modulus changes: II. Shear wave imaging of thermal lesions *IEEE Trans. Ultrason. Ferroelectr. Freq. Control* **58** 1603–11

[25] Cohen R J, Shannon B a., Phillips M, Moorin R E, Wheeler T M and Garrett K L 2008 Central Zone Carcinoma of the Prostate Gland: A Distinct Tumor Type With Poor Prognostic Features *J. Urol.* **179** 1762–7

[26] Bohyun K and Chan K K 2013 Embryology, Anatomy and Congenital Anomalies of the Prostate and Seminal Vesicles *Abdominal Imagin* pp 1797–812

[27] Orescanin M, Member S and Wang Y 2011 3-D FDTD Simulation of Shear Waves for Evaluation of Complex Modulus Imaging *IEEE Trans. Ultrason. Ferroelectr. Freq. Control* **58** 389–98

[28] Jiménez N, Picó R, Camarena F, Redondo J and Roig B 2012 Ultrasonic evaluation of the hydration degree of the orange peel *Postharvest Biol. Technol.* **67** 130–7

[29] Virieux J 1986 P-SV wave propagation in heterogeneous media: Velocity-stress finite-difference method *Geophysics* **51** 889–901

[30] Leutenegger T and Dual J 2004 Non-destructive testing of tubes using a time reverse numerical simulation (TRNS) method *Ultrasonics* **41** 811–22

[31] Liu Q H 1999 Perfectly matched layers for elastic waves in cylindrical and spherical coordinates *J. Acoust. Soc. Am.* **105** 2075–84

[32] Gallego R and Rus G 2004 Identification of cracks and cavities using the topological sensitivity boundary integral equation *Comput. Mech.* **33** 154–63

[33] Rus G, Wooh S C and Gallego R 2007 Processing of ultrasonic array signals for characterizing defects. Part II: Experimental work *IEEE Trans. Ultrason. Ferroelectr. Freq. Control* **54** 2139–45